CRACKING THE
PM INTERVIEW

*How to Land a Product Manager
Job in Technology*

ALSO BY GAYLE LAAKMANN MCDOWELL

CRACKING THE CODING INTERVIEW

150 PROGRAMMING QUESTIONS AND SOLUTIONS

THE GOOGLE RESUME

HOW TO PREPARE FOR A CAREER AND LAND A JOB AT
APPLE, MICROSOFT, GOOGLE, OR ANY TOP TECH COMPANY

CRACKING THE
PM INTERVIEW

*How to Land a Product Manager
Job in Technology*

GAYLE LAAKMANN MCDOWELL
Founder and CEO, CareerCup.com

JACKIE BAVARO
Product Manager, Asana

CareerCup, LLC
Palo Alto, CA

CRACKING THE PM INTERVIEW

Copyright © 2014 by CareerCup, LLC.

Published by CareerCup, LLC, Palo Alto, CA. Version 1.01490281772201.For more information, contact support@careercup.com.

978-0984782819 (ISBN 13)

Table of Contents

Please join us at CrackingThePMInterview.com for
additional resources and information, or to discuss
questions with fellow PMs.

Introduction
Chapter 1

Product management is a strange role.

For many roles, getting from point A to point B is pretty obvious. If you want to get a programming job, you learn how to program. So you go to college for computer science or you learn to write code on your own. Ditto for being an accountant, lawyer, doctor, etc.

If you want to be a product manager, what do you do? There are no schools for product management. There's no formal training. It's also not a role you typically get promoted into, exactly.

How, then, does one land a product manager job? That's what this book is here to teach you.

Jackie and I have worked with countless current and aspiring product managers to help them learn how to get the right experience, how to position themselves appropriately, how to prepare for interviews, and how to ace them.

This book translates these many hours of coaching sessions and conversations into written form.

Why does this matter?

Product management shouldn't be this elusive role, accessible only to those who are lucky (and connected) enough to have someone explain what PMing is all about. Greater accessibility is a good thing for candidates and employers alike.

You, the candidate, are better able to position yourself for a job when you know what to expect. It's not about faking anything; you can actually acquire the experience you need once you know you need it.

When it comes time for interviews, you'll able to prepare for their questions more effectively. You will learn how to describe your unique experiences and most important accomplishments. You'll learn how to tackle problem-solving questions. You'll understand what it means to think about the user. And, finally, you'll solidify your technical skills.

Employers, in turn, get more qualified candidates. More relaxed and better prepared candidates perform in a way that is more consistent with their skillset. They know which accomplishments are most interesting and relevant to employers, and they can drive the conversation toward those things. They can demonstrate their problem-solving skills when they know that's what the question is about. They can learn the bits of knowledge that they need in order to tackle specific questions. This sort of preparation portrays a candidate's skillset closer to how it would be on the job.

Taking the guesswork and randomness out of interviews is a good thing for everyone.

Who are we?

If you flip to the back of the book, you can read our credentials: Google. Microsoft. Apple. Startups. Hiring committee. Oodles of interviews and coaching sessions.

Yadda, yadda, yadda.

That's all very well and good, but that won't tell you who we *really* are and how we got here, writing a book on landing a product management job.

I (Gayle) come from a deep engineering background, but I've also spent a lot of time working with candidates: interviewing them for dev and PM roles, conducting mock interviews, coaching them on how to strengthen their answers, teaching them concepts that they don't understand, and discovering what their goals and passions are.

I learned two things through this. First, I learned how much even good candidates could improve their interview performance, with a bit of help. Second, I learned how little information there was about getting a product management role. Lots of people talked about how to be a good product manager, but few people talked about how to actually break into that field.

Except, of course, for my amazing coauthor, Jackie.

I stumbled across Jackie's blog on Quora, the question and answer site. After following her for a bit, I was struck by a few things. First, she had worked for several of the top companies, so she was good enough to navigate these interviews herself. Second, her advice was to-the-point, actionable, and good. Third, she *cared*. She cared enough to write a blog to help people enter the PM profession.

Making the product management field more accessible is important to both me and Jackie. We believe that everyone benefits from this, and that is why we've written this book.

What now?

This book will not help you "fake" your way into a PM job. It will, however, give you a plan of attack, from the beginning to the end of the process.

We will show you what a product manager is and how the role varies from company to company. We will also describe how different companies think about their hiring process.

We will talk to product managers from a variety of backgrounds to understand how they broke into this field and to get their perspective on what makes a good product manager.

We will talk about the different backgrounds PMs tend to have and how you might make the transition given your background and interests.

We will show you how to write a great resume and cover letter. Some of this will be advice that applies outside of product management, but much of it is advice specific to product managers.

We will advise you on how to get ready for an interview: what sort of research you need to do prior to an interview and how to prepare for questions. We will tackle each of the major interview question types—including behavioral questions, estimation questions, product questions, case questions, and coding questions—and explain to you what an interviewer is looking for and how you master those questions.

We will help you put your best foot forward so you can get in the door.

As you read these pages, reflect on how you can draw out the strong points of your experience in an interview and how you can deliver more powerful responses.

We are here to help. Please join us online at crackingthepminterview.com for additional resources and tips.

And finally, just as software is never "done," we hope this book will never be either. If you have suggestions, feedback, questions or just want to say "hi," don't hesitate to drop us a note: gayleandjackie@careercup.com

Thank you, and good luck!

Gayle L. McDowell
facebook.com/gayle
twitter.com/gayle
technologywoman.com
quora.com/Gayle-Laakmann-McDowell

Jackie Bavaro
facebook.com/jackie.bavaro
twitter.com/jackiebo
pmblog.quora.com
quora.com/Jackie-Bavaro

The Product Manager Role
Chapter 2

What is a PM?

A PM is responsible for making sure that a team ships a great product.

Some people will say that the product manager (sometimes called the program manager or project manager) is like a mini-CEO of their product. That's accurate in some ways, since a PM takes holistic responsibility for the product, from the little details to the big picture. The PM needs to set vision and strategy. The PM defines success and makes decisions.

But in one of the most important ways, the description of product manager as CEO misses the boat: product managers don't have direct authority over the people on their team.

As a PM, you'll need to learn to lead your team without authority, influencing them with your vision and research. Product managers are highly respected at most companies, but not more so than engineers. If you show up and start bossing people around, you'll probably find it hard to get things done. After all, engineers are the ones actually building the product. You need them on your side.

One reason product management is such an appealing career is you get to sit at the intersection of technology, business, and design. You get to wear many hats and learn multiple points of view.

As a product manager, you'll be the advocate for the customer. You'll learn their needs and translate those needs into product goals and features. Then you'll make sure those features are built in a cohesive, well-designed way that actually solves the customer's needs. You'll focus on everything from the big picture to the small details. One day you might brainstorm the three-year vision for your team, while the next day you work through the details of the buttons in a dialog.

Product management is a highly collaborative role. The product manager usually serves as the main liaison between the engineerings and other roles such as design, quality assurance, user research, data analysts, marketing, sales, customer support, business development, legal, content writers, other engineering teams, and the executive team. It's usually the job of the product manager to identify times when one of those teams should be brought in, and to fill in for them if they don't exist.

Functions of a PM

The day-to-day work of a product manager varies over the course of the product life cycle. At the beginning, you'll be figuring out what to build; in the middle you'll help the team make progress; at the end you'll be preparing for the launch.

While the product life cycle varies by company (and sometimes even by team), it usually follows a general pattern of Research & Plan, Design, Implement & Test, and Release. Of course, these frequently overlap and feed back into each other.

Some companies or teams split the product manager role across two people: the more business-focused person and the more engineering-focused person. When companies make this split, they call the engineering-focused person the technical program manager or technical product manager (TPM), and they call the business-focused person the product manager (PM).

When a team has a TPM and a PM, the product manager focuses on Research & Planning and Release, while the technical product manager is more focused on Design and Implement & Test. For example, the product manager will research the market and define the requirements. The TPM will work with the PM to translate those requirements into the specific feature work required, and then facilitate the engineering team as they build it.

Research & Planning

All products and features start with research and planning. This is the time when the PM is starting to think about what to build next. The next idea may come from a customer request, competitive analysis, new technology, user research, the sales or marketing teams, brainstorming, or the big vision for the product.

Depending on the scope of the role, a big part of the product manager's job in this phase is creating or proposing a roadmap. This means figuring out a cohesive long-term plan for the team. The PM talks to all possible sources to create a large list of potential features or development work. Then, based on factors like customer needs, the competitive landscape, business needs, and the team's expertise, he prioritizes the features and scenarios.

Once the PM has a proposed roadmap, he needs to bring other people on board. Some companies, such as Microsoft, Apple, and Amazon, have a top-down approval process, where executives and directors get involved very early. Other companies, such as Google, Facebook, and many startups, have a more bottom-up approach, where the PM focuses on winning over the engineers.

Once he's chosen a feature set, the product manager becomes the expert on them. He'll think deeply about the problems he's trying to solve and the goals of the features. In the coming phases, everyone on the team will have questions, including "why are we working on this?", and the PM will need to have answers.

This is also the time when the PM starts defining success. He'll envision what the world looks like if the team is successful. Many companies use the model of Objectives and Key Results (OKRs) to communicate the most important goals of the team. In this model, the PM works with the team to come up with measurable results that it can commit to.

Design

Once the PM has formed agreement on what the team is going to build, it's time to design the product and features.

Product design does not just mean user interface (UI) design or drawing out what the product will look like. Product design is defining the features and functionality of the product. The PM's role in product design varies substantially between companies and teams.

On some teams, especially shipped software (as opposed to online software) teams at Microsoft, the PM will write out a detailed functional specification (spec) that includes things like:

- Goals

- Use Cases

- Requirements

- Wireframes

- Bullet points describing every possible state of the feature

- Internationalization

- Security

The spec will then spend weeks being inspected, reviewed, and iterated on by developers, testers, and other PMs. On these teams, the PM is expected to make every user-facing decision.

Other teams have much looser specs and a more rapid design process. The PM might sit down with an interaction designer, chat about the goals, brainstorm on a whiteboard, and then iterate by giving feedback on the designer's mock-ups. When the mocks are ready, the PM might send them to the engineers with just a few sentences in an email. On these teams, the engineers will generally make easy product decisions as they come up and call the PM over to ask about the more difficult ones.

And for some teams, especially at Apple, the design is done mostly by a dedicated design team with minimal input from the PM. On those teams, the PM might focus more on project management and fighting fires as they come up.

Since the product manager's role in product design can vary so much across teams, it's a great thing to ask about during your interview. Ask about who you'll be working with on your core and extended team. Find out how much of your time will be spent writing specs and how much you'll be working with designers. Learn where the balance is between PMs, designers, and engineers in making product decisions.

Implement & Test

A PM's work isn't done once engineers start coding. During the implementation stage, the product manager keeps track of how the project is going and makes adjustments.

During implementation, one of the most important parts of the job is helping the engineers work efficiently. The product manager will check in regularly with his team and learn how things are going.

Often an engineer will be blocked because she's waiting for work from another team. In this case, the PM will need to find other tasks for the engineer and, in the meantime, work with the other team to get the work finished more promptly.

Sometimes, implementation of a feature will turn out to be harder than anticipated, and the PM will look for ways to change the feature to make it easier to implement. If an engineer is running behind schedule, the product manager can review the scheduled work carefully and cut lower priority work.

During implementation, the product manager will also start gathering feedback and reporting bugs on the early versions of the product. Sometimes, a feature that looked good during the design phase will not work as well as expected once it's used in the real world. To find problems like this, teams will do usability studies, run experiments, and do internal "dogfooding."

Dogfooding comes from the term "Eat your own dogfood," and simply means using your own product yourself. For example, people at Microsoft run early versions of the next Windows release on their computers every day. Facebook employees use Facebook Groups to communicate.

Sometimes teams need to be more creative to find ways to use their own products. For example, Google gives employees an AdWords budget and encourages people to create advertising campaigns to make sure they get enough dogfooding.

Another way to find out if a product will work before it's launched is through usability studies. In a usability study, participants try out early prototypes of a new product or feature. Usually the participants are given a scenario or goal, and then they'll try to use the prototype to accomplish the goal.

Larger companies usually have a user researcher, who develops and runs the study with some input from the product manager. At smaller companies, the PM might run the studies. In both cases, by watching a handful of studies, PMs can see where people struggle and identify key usability issues.

While dogfooding and usability studies are great for getting qualitative feedback, running experiments is a way to get quantitative feedback when you have online software. In an experiment, the new feature is turned on for a percentage of users (the experiment group), while the rest of users (the control group) keep using the product without the new feature. For online software, it's common for all new features to be launched gradually as an experiment.

In an experiment, you can measure specific metrics for the new feature, such as how many people click a new button you added, as well as overall success metrics like user engagement, retention, and revenue. By comparing the success metrics between the experiment and control groups, you can tell how successful the new feature is.

As feedback comes in through dogfooding, usability studies, and experiments, the PM identifies the most important issues and iterates on the feature design to find better solutions. Prioritization is one of the product manager's most important functions at this point; if the team were to fix every bug and build every new feature idea, the product would never launch. The PM needs to consider all of the new requests and decide if they should be prioritized for the current release or punted to a later time.

Release

When the development process is finished, the product manager needs to make sure the launch goes smoothly. The launch process varies from team to team but usually involves things like:

- **Running through a launch checklist.** There might be final approvals from key stakeholders like Legal or coordination steps with teams like Marketing and Operations.

- **Making sure that the teams who will support the product going forward are prepared.** For a web product this might be the customer service team; for a hardware product it could be the manufacturing team; for a service-based product it could be the operations team.

- **Preparing for all the things that could go wrong.** As the release nears, urgent issues inevitably pop up, and the PM thinks on her feet to fight the fires.

After a successful launch, the PM usually announces the launch to the rest of the company, celebrates with the team, and then prepares to do it all over again. Depending on the team, the PM might continue to support the product after the launch, gathering metrics and iterating on user feedback, or the product might be handed over to another team for operations and maintenance.

How the type of product affects the PM job

The actual work a product manager does can depend a lot on what the product is. Software that ships on a DVD or in an app store is very different from online software that can be updated at any time. Additionally, being a PM on a mature product might be very different from being a PM on a new product.

Shipped Software

Shipped software refers to products like mobile apps that ship in the Apple App Store or software that ships on DVDs. Shipped software is unique because it's hard to update after launching. With a web app, you can always release a new feature and quickly roll it back if there are issues. With shipped software, it's important to get it right the first time.

As a result, shipped-software teams tend to have longer timelines, and the products require more project management and coordination between teams. Specs are more important because features are more formal and need to serve more audiences. User research and internal dogfooding (using early builds of the software) are also very important since you need to know if the product is good before it's released.

PMs who are good at project management and have good communication skills do well working on shipped software. Shipped software can also be great for people who want a good work/life balance, since there aren't usually urgent issues that need to be fixed within hours.

Online Software

In online software, being scrappy is very important. Product updates are relatively easy, so things tend to move quickly. Instead of waiting for the product to be perfect, teams will often launch something, see how it does, and launch again.

Most online-software teams run A/B tests (also called multivariate testing or online experiments). In an A/B test, a new feature is released to a percentage of users. The behaviors of the experiment and control group are later compared to see if the new feature improves the experience.

Because companies in online software collect more data, it's important that these PMs are skilled with data analysis and designing experiments. It's also important to work well under pressure, as servers can fail at any time and PMs often have to make quick decisions.

Consumer Products

In consumer products like social networks, photo sharing apps, and web search, the customers are regular everyday people, just like you, your grandmother, and the engineers. The good thing about working on consumer products is that everyone understands the target customer and use cases. Unfortunately, that's also the bad thing.

On a consumer facing product, the engineers will often have lots of product ideas and won't rely as much on the product manager to come up with features and design. PMs can often act like shepherds and editors, guiding the product to a good endpoint, rather than being the person who makes all the decisions.

Data-driven PMs can do very well working on consumer products because they're able to make a strong case for their proposals, and they often can come up with features that will make a difference to the core metrics the company cares about. Consumer products are also great for people who want to be able to explain what they do to their parents!

B2B Products

In business-to-business (B2B) products like online ads or productivity software, the customer works at another company. For these products, engineers realize that they're not the target audience and tend to rely more on the product manager to understand the customer.

Depending on the team, sometimes PMs on B2B products are responsible for thinking about how product decisions will affect revenue. They'll need to balance features that match their long-term strategy against features that current big customers are clamoring for.

PMs who like doing customer research and market analysis could enjoy working on B2B products. These are also the products where PMs tend to exert the most influence, so they can be a very satisfying place to work.

Early Stage Products

With newer products, such as those about to launch or recently launched, the team is often focused on shipping a minimum viable product (MVP). This is not the time to tackle all challenges, since you don't even know if the product is a good fit in the market and for customers. Instead, you want to answer questions and prove your core value as quickly as possible.

PMs focus on cutting non-essential features to strip the product down to just the essentials. This allows them to launch faster and to begin the process of learning what customers really want (or if they want the product at all). Sometimes, this means launching products that aren't as polished as you would like.

PMs who like excitement and are comfortable with doing things the quick and dirty way do well on early stage products. For PMs on early stage startups, the most exciting thing can be taking a product from a tiny user base to a much larger one.

Mature Products

For mature products, such as market leaders, most of the work will be iterating on the product and trying to improve it. PMs often have feedback from previous versions as to which areas need the most improvement and can focus on them.

As a PM on a mature product, it can be very important to make sure you don't get stuck making small incremental improvements. Often, a mature product's biggest competitor is the last version of that same product. At the same time, mature products often have the luxury of time to make big bets on new ideas.

One of the biggest advantages of working on a mature product is that you already have a huge user base. Every improvement you make will be multiplied to make a very big impact. On the other hand, many companies with mature products become risk averse and won't make daring changes.

PMs who want to work on products used by millions of people would enjoy working on mature products. Mature products are also a great place to learn from the people who were able to make the product succeed.

Top Myths about Product Management

Since product management isn't a very well-known role, there are a lot of misconceptions about who product managers are and what what they do. Here are the top ten myths about product management.

1. Product Managers are Project Managers.

While some product managers have project management as a large part of their job, most do not. Project managers are mostly concerned with timelines and coordination. While they might be responsible for gathering the project require-ments, they don't have much say in identifying and choosing the requirements.

Product managers are responsible for identifying problems and opportunities, picking which ones to go after, and then making sure the team comes up with great solutions, either by thinking of the solution themselves or by working with the designers and engineers. This is why product sense—having the intuition to recognize the difference between a good product and a bad product—is so important for product managers.

2. Product Managers are in Marketing.

This myth is tricky, because with title inconsistencies, sometimes there are marketing roles called "Product Manager." But at companies like Google, Amazon, Twitter, and Facebook, product managers are not in the marketing department. Instead, they're usually in the engineering organization.

Marketing folks focus on getting users into the product, while product managers define what happens once the user is in the product.

For example, a marketing manager might come up with messaging and start a social media campaign, while a product manager comes up with new features and works with the engineers to launch them. While marketing people will talk to product managers about features that would help the messaging or branding, they don't define the details of those features or work with the engineers to build them.

3. You can't become a product manager right out of college.

The word "manager" in the title makes many people think that you need a lot of experience to become a product manager. Also, because product managers make so many decisions that affect the direction of important products, it seems like a senior role.

In fact, many tech companies like Google, Microsoft, Facebook, and Yahoo recruit product managers directly out of college. They've found that passion, intellect, a strong customer focus, and lots of energy can be a winning combination for great PMs. If you want to become a PM, don't think that you have to take a different job first.

4. Product Managers just write specs.

The job of a PM is very different than that of an engineer or a designer. Engineers are expected to deliver working code and designers are expected to deliver wireframes and mocks. For PMs, just delivering a spec isn't enough.

PMs are responsible for seeing the entire project through to a successful completion. Writing a spec is a technique for communicating and moving the project along, but the spec doesn't have intrinsic value. Many PMs communicate ideas without specs, through conversations and drawing ideas on whiteboards. And some PMs fail because they write a spec but don't follow through to make sure the team understood and implemented the ideas.

5. Product managers just set up meetings.

Some people think that a PM's job is just to get the key stakeholders in a room together to make decisions. Good product managers don't just serve as passive conduits of other people's opinions. Instead, PMs research the area and come up with their own point of view and frameworks for making decisions.

PMs do need to meet with the key stakeholders and understand their opinions and priorities, but then they synthesize those perspectives, lay out the tradeoffs, and come up with a recommendation that will satisfy all of the stakeholders. In any meeting or conversation, the PM needs to represent the interests of all the people who are not in the room.

Product managers are able to reduce the number of meetings their teammates need to attend because they're able to represent the team to other groups and find productive ways of communicating that don't require meetings.

6. PMs should build exactly what the customers ask for.

It's great to do customer research and listen to what customers ask for, but it's not enough. Product managers look beyond what customers say to see the hidden needs and deeper goals.

When Oxo, a kitchen utensil company, asked customers what was wrong with their measuring cup, they talked about the cup breaking when they dropped it or its having a slippery handle. But when they watched people use the measuring cup, they saw people pour, then bend down to read the measurements, then pour, then bend, then pour, then bend.

Nobody asked to be able to read the measurements while pouring, but Oxo was able to see the need. They now sell a measuring cup with the measurements at an angle so you can see the lines while pouring.

7. PMs set the dates.

As Nundu, a PM at Google, says, "PMs don't set dates. Engineers set dates." As a PM, you can tell your team what you want them to build, but then they'll tell you how long it will take to build it. If the timeline is too long, you can't just tell them to code faster; it won't work.

Instead, if you have external deadlines you want to hit, you need to make tradeoffs and negotiate. You either need to cut features or find a way to parallelize the work and bring on more people to help out. Sometimes you can be even more clever and find ways to reduce the rest of your engineers' workloads, such as getting them out of unnecessary meetings or having them temporarily spend less time interviewing candidates.

Not trusting the engineers' estimates and promising other teams that the work will be done sooner than the engineers agree to is one of the fastest ways to ruin your relationship with the team.

8. Product Managers are the boss.

Some people try to sell the PM role by saying you're the CEO of your team. In reality, product managers have no direct authority over the team. The team is never obligated to do what the PM says.

Instead, PMs influence without authority, building up credibility with the team, communicating clearly, gathering data and research, and being persuasive to lead the team. Teams follow PMs when they're convinced that their goals align and that the PM will help them better achieve their goals.

It's important for PMs not to try to tell everyone what to do, stepping on designers' or engineers' toes. Designers should be empowered to own the design of the product, and engineers should be empowered to own the technical implementation. PMs need to understand those choices and the impact they have on the overall experience. They should be willing to speak up if they don't align, but PMs shouldn't make the mistake of trying to control those decisions.

9. Ideas are more important than execution.

People who are new to PMing sometimes think that coming up with ideas is the most important part of the job. In practice, execution of an idea is much more important. Many different people on a team can come up with lots of great ideas, but the details are usually the hard part.

As a PM, it's important to take broad ideas and make them tangible and actionable. Product managers need to think about the corner cases and figure out all of the little steps that need to happen to make an idea a reality. Often this involves getting your hands dirty: finding servers to run your code, convincing other teams to prioritize the work you depend on, and using the product consistently to find and iron out all the rough edges.

10. You can say "That's not my job."

While most roles on the team are crisply defined, product managers have a more fluid role. When you're a product manager, your job is anything that isn't being covered by other people.

As a PM, you're responsible for the success or failure of your product, and no job is beneath you. If there's work that no one wants to do, you need to find a way to get it done, even if that means doing it yourself. If you let the work slip by, no one else will make sure it gets caught.

Project Managers and Program Managers

Note: Microsoft has a role called Program Manager that is different from the Program Manager role at most other software companies, and is similar to the Product Manager role described earlier.

There are many roles that are related to product management, and the lines between the roles can be blurry. If you're looking to apply for a project manager or program manager role, see if you can talk to people who have done the role at the companies where you're applying to make sure you understand the nuances of the requirements.

Project managers make sure that their projects get delivered on time and on budget, to the satisfaction of the customer, whoever that may be. Program managers (except for Microsoft program managers) are similar but are usually in charge of a long-running program instead of a series of projects with set end dates.

Software companies often have project managers leading teams that are internal-facing, like infrastructure projects and operations programs, or leading consulting teams that are focused on a single customer. This is in contrast to product managers, who usually lead teams building customer-facing products. However, this isn't a hard and fast rule.

The Job of a Project Manager

Two things are unique about these roles. First, the project manager has a very clear and specific customer or goal.

Project managers working on infrastructure and operations teams have another employee of the company as their customer. The other employee often has a clear and accurate idea of what they want out of the project.

Project managers working with a consulting team serve an individual customer who has already signed a statement of work that includes the details of each project.

In all of these cases, the research, planning, and design stages are fairly straightforward, or already done by someone else. For project managers, product design is not usually a big part of their job, so you may or may not get product design questions.

Also, because the customer is so deeply invested in the outcome of the project, communication and expectation setting are very important. Project managers need to be comfortable reporting on the team's progress. Being detail oriented and closely involved with the team are important so that you can answer questions from the customer and explain when things go wrong.

The second unique part of project and program management is that budgeting and resource management are a big part of the job. Project managers often work in cost centers, so they focus a lot on operational efficiency and reducing costs while keeping quality up to par, unlike product managers who are often shielded from those aspects.

Here are some types of budgeting and resource management work that project managers and program managers do:

• Clarify goals and gather satisfaction metrics.

- Determine the people and skills needed to complete a project.

- Set up project management tools, plans and processes.

- Run status meetings and gather status reports.

- Analyze data to identify opportunities.

- Identify & implement changes to improve efficiency.

- Manage changes that come in from the customer.

- Find ways to keep the project on track even when things go wrong.

Requirements for project managers vary from company to company and role to role. Some teams look for project management certifications, while others do not. Generally, companies look for project manager candidates who have prior experience managing projects and excellent communication skills.

Companies
Chapter 3

One of the biggest questions on a PM candidate's mind is "What's the difference between all of these companies?" While the PM role sounds very similar at many companies, in practice there can be a big difference in what day-to-day life is like.

How the PM Role Varies

We talked to PMs from Amazon, Apple, Facebook, Google, Microsoft, Yahoo, and numerous startups to learn what the product manager role is like at each company and what makes the company unique.

We learned that the companies vary on factors like transparency between divisions, work-life balance, and how much they value candidates with technical background versus business degrees. The PM role also varies in scope, in terms of how much the job involves product definition, design work, strategy, and project management.

Transparency

Some companies, such as Google, Facebook, and Yahoo, are very transparent, with lots of visibility into what other teams are working on. Others, such as Apple and Amazon, are more siloed, with each team focused on their own work.

At the more transparent companies, you'll frequently see PMs move between teams, and cross-team collaboration is a big part of the PM's job. At the more siloed companies, there's less movement and usually less of the high-pressure, cross-team work.

Ratio of PMs to Engineers

The ratio of PMs to engineers can also vary widely. Microsoft has many PMs, with the ratio in some teams as high as 1:3. At other companies, a ratio of 1:10 is more common. Google and Twitter are known for having very few PMs per engineer. The ratio has a big impact on how closely involved a PM is in the day-to-day work of the engineers and how large a product she's responsible for.

At a company with lots of PMs, there is a lot of collaborative work and there are many chances to learn from people with more experience. At a company with few PMs, there is a greater chance for ownership of a large area and for independent work.

Product Strategy

A PM's role in defining product strategy differs between companies. At some companies, the strategy is "bottom up," where key decisions often come from developers and PMs. At other companies, the strategy is more "top down," where the direction is generally decided by executives and PMs, and developers are left to implement it.

At Google, Facebook, Yahoo, and Amazon, product managers are deeply involved in product strategy, deciding which direction to take the product and when to start new initiatives. PMs are expected to think about strategy for their teams, for example, which customers and areas to focus on. They'll get advice and opinions from lots of people, but ultimately they need to present a compelling plan. Often, PMs at these companies can easily get an engineering team to build out ideas at least to the experimentation stage, not needing executive approval for launches until later.

Meanwhile, at Microsoft and Apple, the strategy tends to come from the top down, while individual PMs execute on that strategy. This isn't to say that PMs in top-down companies can't influence strategy, but many PMs won't until they're quite senior. Great PMs at these companies influence strategy by pitching their ideas to executives and winning them over.

Company Culture

Another big difference is the company culture. At companies like Google, Microsoft, Yahoo, and Facebook, there's a lightheartedness during the work day. These companies are proud of their excellent perks such as free food and drinks, or even free massages. And while people may work long hours, they'll always say that the quality of their work is more important than the number of hours they put in.

On the other hand, some companies, such as Apple and Amazon, have a culture where employees are proud of how hard they work. At these companies, it is expected that employees work long hours, and frugality is valued. Employees at these companies are so inspired by the company's mission that they're happy to work weekends or take calls late at night. Building an amazing product isn't meant to be easy.

Who they recruit

Amazon prefers MBAs for the product manager role and doesn't consider a technical background to be critical. This is one of the few companies that don't hire new college graduates into the product manager role. They do, however, accept new graduates as program managers or technical program managers, a related role that skews more towards project management than product design.

Apple hires for both hardware and software EPM (engineering program manager) roles, so it's a good choice for people with a background in electrical engineering or computer science. Apple hires new college graduates and doesn't tend to hire people with MBAs as engineering program managers.

Facebook is the most technical of the group, requiring that all product managers be technical. The company values its entrepreneurial "hacker" culture and has a substantial number of PMs who were founders of acquired companies, as well as quite a few ex-Google PMs. They hire new college graduates into the rotational product manager role, which includes 3 four-month rotations across teams.

Google prefers to hire new college graduates, usually computer science majors, starting them in the associate product manager (APM) program, a two-year rotational program. Some Google PMs have MBAs, but Google places more emphasis on master's degrees or PhDs.

Yahoo mostly hires experienced PMs but has also started an associate product manager program to hire new graduates. Yahoo values people with technical backgrounds who can communicate well with engineers. They look for passionate people with good product sense and who are entrepreneurial, but also have a realistic sense of what it means to deliver software to hundreds of millions of users.

Microsoft hires both new college graduates and experienced hires for the Program Manager role and prefers a technical background, though it doesn't have to be specifically computer science. Separately, Microsoft hires MBAs into the product manager role, which is a marketing role at Microsoft. Microsoft does well at recruiting internationally and hires many PMs from outside the US.

Google

Google's structure reflects its startup roots. Google is passionate about innovation and really values a culture where great ideas can become reality. Google's vision comes from the bottom up, and teams are often engineering driven. Product managers focus on strategy, analysis, and facilitation of the engineering team.

One thing that makes Google unique is the incredible transparency across the organization. Most code and documentation is available to any full-time Googler, and the executive team takes questions from all Googlers at the weekly all-hands "TGIF" presentation. It's not unusual for Google PMs to switch product teams during their career at Google.

Who they recruit

Google looks for entrepreneurial self-motivated people who love technology. Candidates with MBAs or more than four years of experience join as product managers, while those with fewer than four years of experience apply to be associate product managers (APMs).

The APM program is an elite two-year product management training program for new graduates. APMs are given important product management roles on teams across the company, along with training, networking opportunities, a midpoint job rotation, and an international business trip to meet Googlers and customers around the world.

What they do

Google has a lot of products, and the product management role can vary a lot across teams. While product design thinking and analytical skills are important in all of the roles, the balance will vary a lot across products like Google Search, Adwords, Gmail, Android, YouTube, Google Plus, and Google Maps.

The Search team tends to be very research driven, with engineers taking the lead on inventing new algorithms. For advertiser-facing teams, PMs gather customer requirements and communicate those needs to the rest of the team. On teams like Google Plus, designers are central to the team, while developer-facing teams might not have a designer at all.

PMs at Google work independently. PMs join a team, and it's up to the individual PMs and their teams to decide what they're going to build. Often, a PM's first project at Google is to figure out what they should be working on.

Many products have only one PM, and for those that have more, the work is usually cleanly partitioned so that each PM owns a full area. In the day-to-day work as a Google PM, you'll work most closely with your engineering team and designer. Many new ideas come from PMs, engineers, and designers drawing ideas on whiteboards and then quickly building prototypes.

Google strongly values analytical skills in its PMs, since data analysis can be a big part of a PM's job. PMs frequently look through the usage logs to come up with ideas for new projects in the Search and Ads divisions. Once a team has built something, Google makes it easy to try it out in front of a tiny percentage of users. Then the data starts rolling in, and the PM analyzes the data (or works with data analysts) to see if the changes were an improvement.

A big part of the PM job at Google is getting projects in shape for launch. Since even small changes will be seen by many millions of users, it's important to get all of the pieces right, including the UI and algorithms. A project must also meet security, legal, and infrastructure requirements. Many projects go through multiple rounds of iteration before getting a final approval.

Innovation

To encourage innovation, Google has a program called "20% time." This is a policy where engineers and PMs can spend 20% of their time on a company-related side project. To start a 20% project, you don't need any approval; you just start working on it. There are internal sites where you can post your project and try to recruit other people to join you. Many big products at Google such as Gmail, Google News, and Orkut started in someone's 20% time.

All of this innovation and freedom can make Google a PM's dream. If you have a project you're passionate about, not only do you get the time and freedom to work on it, but you're also surrounded by brilliant engineers who have extra time to help out!

Microsoft

Microsoft originated the PM role in the 1980s when they realized that they needed someone between the marketing and engineering teams who focused on making the product usable for customers.

At Microsoft, the program manager role is unique in scope and influence. A PM serves as a business analyst, a project manager, and a creative force. Microsoft also has one of the highest PM-to-developer ratios. These combine to make the Microsoft PM role a very hands-on position. Teams are frequently PM-driven, with the program manager making every user-facing decision.

As one of the older tech companies, Microsoft has developed a strong focus on career growth. They've learned that employees want to see their careers progressing and have built a system where people can grow not only by becoming team leads, but also by becoming responsible for larger products and more product strategy. Microsoft is a company where many employees expect to spend their whole career.

Who they recruit

Microsoft looks for program managers who are big-picture thinkers, who can solve problems, and who can get stuff done.

Uche from Microsoft says, "We wants people with inquisitive minds. People who look at things from multiple perspectives. As a PM, you'll wear many hats, so how you think is more important than any particular technical skill."

Microsoft has two roles related to program management that are usually filled by people with MBAs: product managers and product planners. Product managers at Microsoft are on the marketing team, identifying market opportunities and developing strategies for moving on those opportunities, focused on the current release. Product planners are looking further ahead, identifying market and technology trends to come up with new product scenarios.

What they do

On Microsoft products, the vision and strategy often come from the top and work their way down.

For example, the VP in charge of Microsoft Office will work with the group program managers to create and share a vision document with the main areas of focus for the next release. The people in charge of each Office product will define the vision for their product so it falls in line with the overall Office vision. Then each of the team leads uses the Office vision and the product vision to develop visions for their feature areas.

The result is that the product strategy at Microsoft within a division is very cohesive. The teams feel like they're working together towards the same goal, and it's rare to find two teams that are working on competing features. As your career advances at Microsoft, you pick up larger and larger pieces of the strategy.

The top-down vision makes it hard to make big changes in direction in the middle of a launch cycle. Even if the idea is great, it's hard to find developers with spare time to build it. On the other hand, you're generally working on features that everyone agrees are important. This means that you can spend your energy trying to build something great instead of convincing management to let you launch the great thing you built.

As a new PM at Microsoft, you'll be put in charge of feature areas that work with the team's vision, and you will be given a lot of freedom and responsibility to make those parts of the product great. As you prove yourself on your early teams, you're given more and more responsibility.

The core feature team at Microsoft includes a developer, a program manager, and a tester. Recently, some teams are bringing a designer into the core feature team. Together the core feature team, led by the PM, decides what to build. On many teams, the PM starts by writing a one-page "spec"—a high-level description of the goals and use cases.

After reviewing the one-page spec with other program managers on the product, the PM may expand the one-page spec into a detailed spec that describes exactly how the feature will work, from high-level flows down to the text of the error messages. The PM might also work with a designer or may do all of the interaction design herself. This is team dependent though.

Once the spec is reviewed and implementation starts, dogfooding (trying early builds of the software internally) becomes very important. Especially on teams with slow ship cycles, PMs at Microsoft gather feedback from other people at the company. As that feedback comes in, the PM prioritizes the bugs and new feature ideas.

Outside of the core feature work, PMs will also take on some team-wide or cross-team responsibilities. One example is running the project management schedule for the product or running a pre-launch triage process to decide which bugs to fix or which features to implement.

As Microsoft adjusts to its new reorganization as a devices and services company, the spec-heavy process is being replaced with a more agile and iterative approach on many teams. Teams are starting to launch faster and more frequently. A/B testing is becoming more prominent.

Apple

Apple has a top-down, siloed structure. The product direction is tightly controlled by the executive team and designers, while the rest of the company executes that vision like a well-oiled machine. At Apple, engineering project managers and engineering program managers (EPMs) are the leaders of the product who keep that machine running.

Who they recruit

Apple looks for people who live and breathe Apple products. While many companies take pride in supporting a healthy work/life balance, Apple looks for people so passionate about the end product that it is their life.

For the EPM role, they're generally looking for someone with a strong background in science and math (so that he can "figure things out") and with a good demeanor (so that he comes across appropriately confident). Also, they want someone who's technically fluent enough to be part of the discussion, but who won't necessarily try to do any of the engineers' job for them.

Since Apple has many hardware projects, they hire for hardware EPMs and system EPMs, in addition to software EPMs. Software EPMs generally have a background in computer science, while hardware EPMs may have a background in another engineering field like electrical engineering or mechanical engineering.

EPMs at Apple come from all levels of experience, from new college graduates to people with 15 years of industry experience. Most EPMs come to Apple from engineering school or engineering roles rather than from business/management degrees, MBAs, or former EPM positions at other companies.

Apple also has a product marketing manager role, which they sometimes call product manager. For product marketing manager, they look for people with a background in business or marketing. MBA graduates are often hired into this role.

What they do

Early ideas at Apple can come from the top down or from the bottom up and are shaped over a series of reviews with the executive team or senior management, driven by EPMs. Product managers do customer research and look at market trends to identify strategic areas for the next release. Once the product is approved, the EPM leads the engineering teams to build the product, creating development schedules, facilitating cross-functional communication, and spearheading issues as they arise.

Products at Apple involve coordination across many teams, and the EPMs are the hub of that communication, making sure things are running on schedule and solving problems when they're not. For example, building a hardware device involves coordinating work across mechanical design teams, electrical design teams, external contract manufacturers, and operations teams.

A typical day for an EPM in Cupertino involves a lot of communication: presenting the team's progress, learning about the status of other teams, and reviewing the current status with executives. Throughout these discussions, the EPM is responsible for surfacing issues and finding ways to resolve them.

For software EPMs, a big part of the day is testing the daily builds, finding blocking issues, and making sure they get fixed in a timely manner.

System EPMs focus on delivering the whole product. They plan and lead prototype builds that dozens of Apple engineers travel overseas for. In addition to being the face of the Apple engineering team at contract manufacturers, the system EPM is responsible for cooperatively driving failure analysis to closure, reconciling development hardware demands, and removing obstacles that block prototype builds. The system EPM is the leader of the project within Apple and pulls together other EPMs and engineers to deliver a product.

Hardware EPMs focus on the hardware deliverables for a product, including circuit boards and flexes. They work hand in hand with electrical engineering teams, product design teams, silicon teams, and external vendors. Day to day, they are involved in silicon planning and supply-chain management, and they are often tasked with reaching consensus on difficult cross-functional design decisions.

Facebook

Facebook is a scrappy, engineering-focused company. There are relatively few product managers; many teams start without a product manager, only bringing one in after the need is obvious. Even when a team has more than one PM, the areas will be big enough that each person is working independently.

Who they recruit

Facebook has a unique set of PMs.

Facebook looks for highly technical and entrepreneurial PMs. At Facebook, all product managers are expected to code (or at least learn the basics) and go through Facebook Bootcamp, a six-week program where PMs and engineers learn the tools and fix bugs. This fits into the do-it-yourself culture; PMs will often code up initial prototypes of their product on their own.

When Facebook acquires a company for the purpose of recruiting its employees, it's referred to as acqui-hiring. When acqui-hiring a team, Facebook looks for small teams, usually fewer than 10 people, composed mostly of engineers. Often, the founder or CEO of the company will be brought on as a product manager.

New graduates and people without product management experience join the rotational Product Manager Program. This is a one-year program consisting of 3 four-month rotations on different teams.

What they do

At Facebook, there are some planned company initiatives, but many ideas spring up from watching how people use the site and the problems they hit. PMs will notice an area that needs attention and put together a storyboard and proposal for what they want to build, including the expected outcome. Because PMs have so much freedom, it's important to know how to position your ideas into the framework of the company and show how they fit into the overall mission.

Then, the team will build a prototype to test in a small market and see if the expected outcome happens. If all goes well, things go into motion. The PM will review the proposal with Mark Zuckerberg (whom everyone calls "Zuck") or the division head to get approval. Then the team brings in designers and turns the feature on internally to start getting feedback from other people at Facebook.

During development, the product manager iterates through product reviews. There's not much project management overhead, but PMs can watch the logs to see new code as soon as the engineers check it in. When everything is ready, the PM coordinates a rollout plan and works with marketing for the launch.

Amazon

Amazon's culture is guided by their 14 Leadership Principles (see: Amazon Leadership Principles, pg 358). While many companies may have their own set of principles, Amazon's Leadership Principles play a real role both in hiring and in day-to-day work. In meetings, people will cite leadership principles to help resolve decisions, and interviewers will look for these qualities in candidates.

Who they recruit

Amazon has a lot going on. There are new initiatives being started, technical operational programs that are ongoing, and non-technical operational programs. There are teams that focus on audiences as diverse as consumers, publishers, and developers. Because of this, there are many different roles that relate to product and program management.

Product managers are the product owners. They focus on the vision of a product. Amazon prefers MBAs for the product manager role and will hire people right out of business school. Unlike many other companies, Amazon does not require PMs to have a technical background.

Technical program managers are responsible for the day-to-day execution of technical projects. TPMs need a strong technical background and can come directly out of school or by transferring from an engineering role. TPMs work closely with engineers.

Program managers are responsible for the project management of non-technical projects, such as those on Operations. Program managers can come from any background, but skill with SQL is a plus. Amazon looks for program managers who are smart, fast, and can work under pressure. Program managers are responsible for running and improving the team's processes.

What they Do

Product managers own the vision and roadmap for their team. Amazon values customer obsession, and many new ideas will come from figuring out what customers want and need. This can be from talking to customers directly or by looking at data that's available via SQL. Amazon is very data driven, so PMs need to have strong analytical skills.

When a PM comes up with an idea, he puts together a business case in a memo, also called a narrative. This document will cover the details of the recommendation and analysis that supports it, especially including numbers about the impact and rationale. Amazon focuses on documents instead of presentations for new proposals because documents force the author to be precise and show clear thinking.

After initial rounds of revisions and redlines, the proposal is then shared with upper management in a meeting that starts with everyone silently reading the document at the same time. This process might seem strange at first, but it guarantees that everyone has time to read the details of the proposal and gets everyone focused. After everyone has read the document, people ask questions about it.

The proposal may then go through a few more rounds; the key is precision in drawing a line from inputs to expected outcomes. After the proposal has the green light, the team starts building it. Many teams follow an Agile process, where the PM is the product owner and is responsible for writing user stories and the team's backlog.

When products are ready for launch, the PM works with the marketing team and prepares for a hand off to the Operations team that will run the program on an ongoing basis.

Yahoo

Yahoo has a strong legacy: at one point, "Yahoo" was synonymous with "The Internet" for many people. The company has gone through a series of transitions in the last several years, but in 2012, Marissa Mayer was appointed as CEO, and she brought renewed energy and passion into the company. Web traffic is up, many recent product announcements and launches have been well received, and you can see a clear change in perception both internally and externally.

PMs who work at Yahoo love the excitement of having such a clear challenge. They're now focused on getting back into the hearts and minds of users by putting their needs first. With a former product manager at the helm, Yahoo is strongly product driven and looking to build products that users love.

Who they recruit

PMs at Yahoo generally have a CS degree, although it's not required. Yahoo looks for talent with strong engineering background demonstrated through academic achievements and/or work experience at a top-tier software company.

Today, Yahoo has a great mix of PMs: some have been at Yahoo for a long time, others have joined recently from other tech companies, and still others are fresh out of college. While each Yahoo PM possesses a different background and experience, they all have one common goal of bringing Yahoo products and its experiences to the next level.

Yahoo started the Associate Product Manager (APM) Program with a goal of hiring talented new graduates. It's a two-year rotational program with regular training presentations from people across the company and a global business trip in the middle. Two things that make the Yahoo APM program stand out are that APMs learn which team they're on before they start working and that the organizers are focusing on getting all of the APMs into big, challenging roles.

What they do

Teams at Yahoo are often organized into a trio of engineer, product manager, and designer, with one person serving as the group lead. The group lead is oftentimes a product manager.

The product manager is responsible for setting the overall direction and strategy for products, making sure the user experience is solid and developing long-term growth. This includes the user-experience and long-term monetization plans, and teams are no longer pressured to look for short-term revenue wins.

Yahoo has a strong culture of collaboration, and ideas come from every corner of the company. Getting the mobile experience right for Yahoo's core products is a key driver of the company's efforts.

Most Yahoo PMs are very technical, and PMs are well respected within the company. PMs are involved in the full spectrum of the product life cycle; they run experiments, work closely with cross-functional teams, and oversee the entire launch process. There is an official review process for product launches that helps keep the quality bar of the products high before they are made available to users.

Teams at Yahoo are moving a lot faster, and many are picking up Agile processes. Many teams work in short iterations. Instead of long, detailed specs, PMs now might just write a one-sentence user story, and the engineers will make decisions and get it out.

Twitter

Twitter is a rapidly growing company working on the future of communication. Since Twitter is growing so fast in terms of users and headcount, the organization is constantly changing and evolving. This growth also opens a lot of opportunities for people to move around and grow their roles.

Who they recruit

Twitter hires people who are passionate about Twitter. They look for people who love the product, the company, or the mission.

Twitter hires for two related roles: product manager (PM) and technical program manager (TPM). Product managers usually work on customer-facing teams. Technical program managers usually work on platform or infrastructure initiatives, often across multiple teams. The PM role includes more product design, while the TPM role involves more project management.

For the technical program manager role, Twitter looks for people with a technical background and experience in software delivery. In addition, they look for a "get things done" mentality and excellent communication skills. Many TPMs come with program management experience, but many also come from backgrounds such as consulting, engineering management, or technical architecture.

For the product manager role, Twitter looks for people with a single-minded focus on the user and who are flexible enough to handle the fast pace. Because Twitter has relatively few PMs per engineer, they usually hire more experienced candidates who can keep up with the work.

What they do

Product managers at Twitter wear a bunch of hats. Teams are made up of product, engineering, and design. PMs are involved in concepts and roadmapping as well as finding bugs, prioritizing features, executing on current projects, and thinking about the future. The PM serves as the interface between the feature area and the rest of the company.

The technical program manager role is relatively new at Twitter and in some ways is still being defined. There are several different types of TPMs, and Twitter is building career paths for all of them.

Some TPMs are responsible for release management: shepherding the web and mobile releases, making sure the builds get in, running the dogfooding process (where Twitter employees try out the new builds), and making sure bugs get fixed.

Other TPMs manage broad programs across many teams with hundreds of people including external people across companies. Other TPMs are thought leaders on cross-team topics like the Agile process. Still other TPMs are embedded in a single platform or infrastructure team.

Ideas at Twitter come from all over the company: management, PMs, engineers, designers, and people in other roles like user services. One way Twitter supports this is with a quarterly hack week where people can work on whatever they want as long as it's related to Twitter. Many great ideas come out of hack week, and people can use the time to build buzz around their ideas.

Planning at Twitter happens quarterly, and teams are responsible for their own goals (called "gulls," since Twitter likes to name things after birds). During implementation, teams constantly dogfood (try new features out internally) and present what they're working on through weekly meetings.

The company runs many A/B tests and is data driven. However, it is also careful about being too data driven. Teams use the metrics to understand how things are performing against success metrics and then make changes as necessary, but they still make sure they move quickly and keep the spirit of the product intact.

Startups

Startups can be great places for PMs to make an impact and have a part in shaping their own role. More so than at larger companies, product managers at startups really have to wear a lot of hats and find scrappy ways to get things done.

One of the best ways to get signal on the culture at a startup is to look at where the founders, PMs, and early employees came from. Since product management doesn't have a single, well-known definition, teams generally bring along the definition that they learned from their past companies. A company founded by ex-Apple employees will tend to have elements of Apple's culture, while one founded by ex-Google employees will tend to be more similar to Google.

Once you're ready to learn more about a startup, it can also make a lot of sense to reach out and try to get coffee with someone on the product team. At a startup, all of the employees are often involved in recruiting, so they'll generally be happy to chat. Plus, they work at a startup, so they probably need that coffee!

Who they recruit

Startups vary a lot in the profile and skills they look for. In general, they're more likely to look for people with experience shipping software and less likely to consider new graduates unless the company already has a large PM team.

For startups, culture fit and passion for the company are really important. While a big company might be satisfied that your reason for wanting to join the company is "I want to work with a great team and have a big impact," that's not going to fly at a startup. They'll want to hear a much more specific story about

Getting the job

If you're looking to land an early PM job at a startup, making a strong impression with your vision for the future of their industry can help you. Shalendra Chhabra (Shalen), now director of marketing at Indix, was able to get hired as an early PM at two different startups by using this approach, despite lacking strong connections to the companies.

When he was looking for a new job, he brainstormed a list of startups with a friend. Swype, a company that makes on-screen keyboards for touch devices, was on that list. He was interested in the company and wrote a memo about how he saw the future: touch devices would explode in popularity, and they would want to expand into different languages and add new features such as predictive tap.

In the memo, he described how his background and skills would add value to the company. He then was able to get an introduction to someone at the company through a friend of a friend and sent along his memo.

Writing the memo took time and effort, but it was worth it. It helped him stand out to the company and gave him a chance to make a first impression on his own terms. When he walked into his interviews, he did so with credibility since people had circulated the memo. Eventually, Swype had a very successful exit, being acquired by Nuance for $102.5 million.

why you're interested in their space.

Referrals are important at all companies, but especially so at startups because they are less likely to want to take a risk. Where there are only a few PMs, each one makes a giant impact on the organization. It can help to reach out to people you know at the company, even if you don't know them super well.

In addition to referrals from employees, startups also look for referrals from their investors, advisors, and board members. You can consider applying through a venture capital firm or via a site like AngelList to get access to all of their portfolio companies.

Finally, many startups recruit their first PM from inside the company.

For example, Paul was in technical support at TechExcel and used his connection with the customers to make helpful product suggestions. The CEO then asked him to lead features.

Similarly, Jesse was an engineer at Venmo who was interested in usability testing. He had shown leadership in organizing the team during a redesign, and the founders asked him to become their first PM.

What they do

Being a PM at a startup can be similar to being a PM at a large company. You'll still be working with engineers and designers, and you're still responsible for the success of your product. PMs at startups help define the roadmap, spec features, triage bugs, analyze experiments, and shepherd launches.

The big difference in being a PM at a startup comes from the scale. Since startups don't have large management structures, the PMs naturally become important leaders for the company. Additionally, startups have fewer resources, so there's more "white space" for the PMs to fill in and more of a need to be really scrappy.

PMs at startups often have a lot of influence over the company culture, processes, and roadmap. Jesse, the first PM at Venmo, was even able to shape the vision statement.

> When I first became a PM, I recognized the need to have a team mission to help everyone stay aligned. I organized a day where we all sat together and wrote down ideas to come up with the one sentence vision for Venmo: "To connect the world and empower people through payments."

For PMs at startups, defining the role is an important part of the job. There's less guidance and mentorship, since you're not coming into a team of experienced PMs. Also, many roles that exist at a big company might not exist at a startup, so the PM might need to take on tasks like customer support, user research, data analysis, or sales until the company grows larger. At a startup, it's unlikely that the CEO will tell the PM to take on these roles; usually the PM notices that there's a gap and starts doing it himself.

One thing that can be difficult as a company's first PM is introducing process. Engineers at startups love shipping code quickly and are very wary of overhead, so it's important to be careful when adding process like Agile. PMs need to make sure that everyone still has a voice and understands the reasons for adding timelines and milestones. It's not always so bad though. In many cases, the team only brings in a PM once it is clear that more process is needed, so the team might already be on board.

While startups have fewer resources within the company, they're often more willing than large companies to go outside the company for help. PMs at startups often reach out to PMs at other startups to get and share advice. Many startups form relationships with other startups, so there's a real community. For example, a B2B startup might find its early customers through connections made by their investors to other startups.

Another big difference is around career advancement. At big companies, PMs are often thinking about getting promoted and climbing the career ladder. At a startup, the focus is much more on making the whole company successful: getting revenue, getting to profitability, having an IPO, or getting acquired. This means that PMs need to take a more holistic view of the company. It's important to make the company, not just your team, successful.

Asana

Asana is an enterprise startup that builds modern productivity software. Unlike traditional enterprise companies, Asana's business model is bottom up: end users bring Asana into a team, not the IT department. This means PMs focus on building software that people love to use. The product team has PMs with backgrounds at Facebook, Google, and Microsoft.

Jennifer, an Asana PM and former Microsoft PM, talks about what it's like to move from a big company to a startup:

> *Because we're a small company with a flat hierarchy, I feel like I got promoted three times when I joined Asana. In addition to PMing meaty areas of the product, I influence other aspects of the company, such as product strategy, process, and culture.*

For example, I get to influence the overall schedule and cadence of the company, what the product's design principles are, how much effort the company puts into creating new features vs. polishing existing features vs. improving internal efficiencies.

And because our team is small and values continuous personal growth, I have the opportunity to work on very different areas of the product. Sometimes I work on features that require a lot of deep UI thinking, and sometimes I work on areas that require more data analysis.

Foursquare

Foursquare is a location-based social networking service that helps people make the most of where they are, whether that's searching for the perfect place for dessert, learning about a concert in Central Park that's trending, or discovering that a friend from out-of-town just landed at the airport.

The PMs at Foursquare come predominantly from Google. Noah, a Foursquare PM, explains the impact this has.

This has certainly influenced our culture. Rather than wholesale import of Google's practices though, we've selectively chosen what makes sense for our size company. Over time, that has changed.

When you're 30 people, individual OKRs [Objectives and Key Results] are superfluous; everyone's would just read 'Do whatever I can to help the company survive and grow.'

Now that we're 150 people, individual OKRs still seem like unneeded overhead, but we've started using team-level OKRs to keep track of firm metric goals.

Compared to Google, we actually write more specs. I think it's because projects move so much faster here. It means we need to quickly and early on work through the biggest and most contentious problems; we don't have time for those to pop up later and cause delays. Specs aren't meant to be comprehensive guides for how to build every piece of a feature. Instead, they're meant to document the non-obvious decisions the team makes so people can reference back to it later.

Dropbox

Dropbox helps over a hundred million people keep their important files safe and available across many devices. You can easily upload photos from your phone, share files, or collaborate on a project. The product team includes PMs from

Google, Facebook, Microsoft, and Zynga, and several were startup founders prior to joining Dropbox.

Matt, a PM at Dropbox, talks about what it's like to work there:

> Working at a 200 person startup is pretty different from a larger company. As a PM at Google, I was very focused on building a great product and making my team successful. But at Dropbox's size, I get to have more impact on shaping the company's culture, how we build products, and how we think about the PM role itself. I get to work closely with the founders on defining our company vision and priorities and building out a great team.
>
> We look for PMs with a technical background, exceptional product instincts, and an eye for detail. Arash, our founder, will notice if a product detail is even one pixel off, so we expect our PMs to really 'sweat the details.'
>
> One of my favorite experiences was working to launch a new security feature (two-step verification) one week after joining the company. We assembled a great team and went from idea to designs to launch in five weeks across all of our desktop and mobile apps, our website, and our developer APIs. It was really gratifying to see how much a small team could do really quickly.

Uber

Uber is evolving the way the world moves. By seamlessly connecting riders to drivers through their apps, they make cities more accessible, opening up more possibilities for riders and more business for drivers. The product team's experience runs the gamut of investment banking, management consulting, startups, and working at larger companies such as Amazon and Google.

Mina, an Uber PM, talks about what makes PMing at Uber great.

> Being an Uber PM is like lifting up the hood of a luxury car. As a user, you just get to enjoy it and you don't need to think about the mechanics that make it work so smoothly. But, as a PM, you get to lift up the hood and retool the complexity to make it work even better.
>
> Users can focus on our apps, which is exactly what we want them to do. In contrast, we look across the whole system from what it takes to find you a car to helping our drivers build their businesses. Because we're so tightly connected to not only our engineers and designers, but also our operational teams, we enjoy a unique role.

On a day-to-day basis, one thing that stands out as reflective of our culture is our global perspective. After only three years, we're in over 40 cities and expanding each week to more cities, which means that every single product we build needs to work internationally. We're also big proponents of workcation - hanging out with your coworkers while building great products in exotic locations, Uber-sponsored. For example, last New Years, we had teams in Melbourne, Miami, Stockholm and Bali ringing in 2013.

Airbnb

Airbnb is a community marketplace for people to list, discover, and book unique accommodations around the world. Airbnb connects people to unique travel experiences, at any price point, in more than 33,000 cities and 192 countries.

At Airbnb, product managers are called producers, like a movie producer. As a producer at Airbnb explained:

Think about a movie. As a producer you work with your creative team and the actors and camera and light to hold it all together. At the end of the day, along with the director, you're responsible for the movie.

The team of producers is very diverse - along with PMs from the big tech companies, there are also people who came from small startups and people who came from the hospitality industry. This diversity enables Airbnb to build a unique culture. The culture is fundamentally design driven and puts a big emphasis on customer research, called "User Insights."

One thing that really makes Airbnb unique is that it's a two-sided marketplace (guests and hosts), and a lot of the experience is offline. Producers at Airbnb think about the offline "frames" (another movie-industry reference).

Thomas, a producer at Airbnb, shared an example:

You might assume that Airbnb has no impact on how kind and professional our hosts are to their guests since this is part of the travel experience that happens offline. In fact, there is a lot that can be done to help hosts to be exceptional hosts who anticipate their guests' needs which then results in meaningful trips, unique experiences and, often times, lasting connections between hosts and guests.

Getting the Right Experience
Chapter 4

If you ask interviewers what they're looking for in PM candidates, they'll usually say that they are looking for smart people who get stuff done.

This desire will be reflected in the job descriptions for product management job openings. The requirements list will be more detailed, but it will ultimately boil down to two criteria:

- Can you be trusted to make the right decisions?

- Can you push through all of the potential roadblocks to deliver a great product?

You will want to focus on these criteria when you're thinking about what kinds of experience to acquire.

For example, whenever possible, see your projects through to the end. Focus on understanding the things you did and the outcomes of your actions. Understand not only if your outcome was successful, but by what metrics it was successful— or not. Consider what it is that drove the success or lack of success. It's okay to fail sometimes, but you need to know why you failed.

New Grads

Product management is a great role to get into right out of college. Many of the big companies have university recruiting programs, and they pride themselves on training new grads to become top notch product managers. In many ways, it's easier to get into product management as a new grad than as an experienced candidate.

The new grad product management programs only accept a small percentage of the many applicants they get each year. If you want to stand out from the crowd, you will need to show strong technical skills combined with excellent customer focus and product design skills. Generally, it won't be enough to just ace all of your engineering classes, since that might still land your resume in the Software Engineer pile.

If you're a current student and want to improve your chances of getting hired as a product manager, consider the following:

- **Major in computer science, or at least get a minor in computer science or a closely related field.** College recruiters often only consider candidates with a very technical background.

- **Pick up a double major, especially in a field like economics or business.** A computer science / economics double major is a very common background for product managers because it shows an interest in both the technical and business side of software. Econ is also a great field for picking up analytical skills and proficiency with statistics. Other fields like psychology, philosophy, cognitive science, human-computer interaction (HCI), or sociology can also be very relevant for product management.

- **Take group project courses.** Group project courses are a great way to pick up leadership skills and start to gain relevant experience. Pay attention to the group dynamics and the challenges your group overcomes. You can use those experiences as anecdotes during your interview to answer behavioral questions.

- **Take on a leadership role.** This can be on anything from a sports team to a club to class president. As a product manager you'll need very strong leadership skills to lead a team of engineers. It's especially great if you can do

Should you intern as a developer or a PM?

You don't have to be an engineer before you can apply to be a product manager. But if you love coding and want to buff up your technical credentials, an engineering internship gives you a great chance to see the PM-Dev relationship from the other side. Many engineers love working with PMs who used to be engineers.

something from scratch, like launch a new club, a school-wide contest, or an extracurricular activity. This shows initiative.

- **Start a side project.** One of the best ways to rise above the crowd is to have a side project like a mobile app. This gives you a chance to show your customer focus and product design skills. If you don't have the technical skills to do this yourself, you can learn them, hire some developers to do the building, or partner with your technical friends.

- **Intern as a product manager or software engineer.** There's nothing better than learning on the job, and an internship gives you that chance.

The key theme here is to show something beyond coding skills. Coding skills are great—often required, in fact—but they're not sufficient. Find a way to show leadership, business skills, and initiative.

Making the Most of Career Fairs

If you don't look like a traditional PM candidate, career fairs can be a great way for you to get your foot in the door, especially at smaller companies. When you chat with the people at a career fair booth and hand over your resume, they'll make some notes based on the impression you made. Those notes can mean the difference between getting an interview or having your resume ignored.

Here are some tips to make the most of your time at career fairs:

- Research which companies will be at the career fair and decide which ones you are the most interested in. Check if they have PM roles available.

- Pick out your best talking points. Think about what will make you stand out and look like a great PM candidate. Maybe it's a challenging class project you took on, or maybe you had a great experience as a teaching assistant. Anything you've done to show initiative (Did you built an iPhone app for fun? Did you start a new competition in your city?) would be a great thing to discuss.

- Practice a short intro that includes your talking points. Imagine walking up to the booth, saying hello, and introducing yourself in a way that will let you talk about your accomplishments in a non-awkward way.

- Think of good questions to ask the company employees. You might not know ahead of time if the company is sending PMs, engineers, or just recruiters, so make sure you have some appropriate questions for each.

- At the career fair, pay attention to how crowded the booths are. This can be a great time to get to know a smaller company and consider places you hadn't thought of before. If the booth isn't crowded, they'll probably be willing to spend longer talking to you, which means more time for you to convince them you're a good fit.

- When you go up to a booth, let them know that you're interested in the PM role, and ask who is the best person to talk to. They might be able to direct you to a PM, or at least to someone who is comfortable evaluating PMs.

- Have a friendly conversation with the employee. Remember that at a career fair they're trying to sell you on the company, so it's entirely appropriate to ask questions. Tell them about why you're interested in being a PM and why you think you'd be a good fit. Show interest in their company.

- Don't just hand over your resume without talking to people at the booth first. If you do, you're missing out on the unique benefits of being able to talk to company employees firsthand.

While you do want to remain professional, don't be afraid to do something a little different. It's good to have some personality and passion! If you've built something particularly cool, you could show the booth staff some pictures or a very short demo.

Do you need an MBA?

An MBA isn't a requirement for product management, and at some startups it might even count against you. On the other hand, teams with more of a business focus consider an MBA a real asset, and some companies, such as Amazon, explicitly focus on hiring MBAs.

Arjun, who got an MBA after starting as a PM at Microsoft, decided to go to business school when he noticed that well-designed products didn't always become market winners. "I saw that good design isn't enough," Arjun noted. "There's something else, and I wanted to learn what that something else is. In business school I realized you could ask business questions and answer them really easily, on the back of a napkin."

This analytical approach to products changed how Arjun thought about features and prioritization. Instead of guessing what users would want, he started thinking about which metrics he wanted to drive.

Here are some tips on how to get the most out of your time in business school if you're interested in product management:

- Take the chance to start something. Gain experience by launching projects, joining clubs, or building something. This will help you avoid the biggest MBA pitfall: being someone who only wants to tell people what to do, and doesn't know how to actually do things.

- Take project-based classes where you can work on your ideas. This lets you make double use of class time, and you have the added benefit: a team of MBAs helping you out.

- Practice designing products by sharing your mockups with classmates and iterating based on their feedback.

- Choose relevant classes like entrepreneurship, marketing, or consumer behavior.

Chris, who got an MBA after being an engineering program manager (EPM), found a lot of benefits from business school. "An MBA gets you experience, connections, ideas, resources, seed funding, and partnerships. Business schools provide a lot of resources," he said.

Does your online persona matter?

It's pretty common for people to have a big online presence these days. From Twitter to Quora to personal websites, there are a lot of ways to get your ideas out there. So should you spend a lot of effort building your website or tweeting?

There are two main places where your online persona will come into play: recruiting and startups. Recruiters often search online to find qualified candidates and may reach out to you to apply based on your online content. If you're hoping to get noticed, being active online can pay off.

LinkedIn is one of the most valuable sites to focus on for recruiting. One way to optimize your profile is to look at PMs with jobs you're interested in and see what their profiles look like and how they stood out. You can also pay for a premium account to see how people are currently finding your profile and optimize around that.

If you're applying to startups, you might also find that some of your interviewers will look you up and notice your online content.

However, at larger companies it will be much rarer that interviewers will research you in advance and discover your website or posts. If you do have a great website, you can include it on your resume or try to bring it up during your interviews. But otherwise, don't worry about it!

Current PMs

If you're a current product manager, moving to the same role at another company is pretty straightforward. If you've already worked as a product manager at a well-known software company, that experience is usually enough to get you an interview at another. There's still room to build up your experience though:

- Launch! The most important way a product manager is judged is by the products she's launched. If your team is close to launch but not quite there, you might want to wait until the product is fully launched to start applying. Likewise, if you're on a team with a very long ship cycle, you might consider switching to a team with more frequent launches so you can get the experience of seeing a product through the entire cycle.

- File for patents. While software patents are a very controversial topic, many companies still see them as a necessary evil. If you work for a company that files patents, make sure you file patents for your innovations. A patent application is a great way to make your resume stand out.

- Take on responsibilities to round out your skills. If you've always been really strong in product design, see if you can learn data analysis. If you've been working on deep technical problems, see if you can spend some time doing user research.

If you're trying to move from a less prestigious company to a more prestigious one, these tips can be especially useful.

Why Technical Experience Matters

Many product management roles list a requirement for a degree in computer science. At first this might seem baffling: coding isn't a regular part of a product manager's job, so why don't companies loosen up that requirement to find someone who truly excels at the core product management skills?

Here's the simple answer: many people without a background in computer science struggle to form a strong working relationship with engineers.

All of those excellent product management skills will go to waste if the product manager alienates his engineers and can't earn their respect. Product management is a job where you have to lead without authority. The only way to get great work done is to bring the team onboard with your vision.

That said, a computer science degree isn't a magic bullet for forming great relationships with engineers, and it's possible to be a great product manager without a technical background. Companies tend to use technical experience as a proxy for the real qualities they're looking for:

- **Able to form a relationship of mutual respect with engineers.** Companies almost always hire a product manager to join a team of engineers who already work for the company. They're not willing to hire someone who won't get along with the team or who can't earn the respect of the team.

- **Good intuition on how long engineering work should take.** A good product manager understands the technical framework he's working with and can help the team prioritize and make tradeoffs between the time spent on engineering and the value of that work to the customer.

- **Scrappy and Self-Sufficient.** Great product managers are action-oriented and passionate about delivering results. They will try to take care of what they can themselves, whether that's gathering data or fixing typos in the product. This frees up developers from the more tedious tasks so they'll be able to do more valuable work.

If you don't have this technical background (or it doesn't come out in your resume), try to find a way to develop and demonstrate these skills.

Transitioning from Engineer to Product Manager

With all of this emphasis on technical experience, engineers are in a great place to break into product management. As an engineer, you really understand what it takes to build a product and the impact that various tradeoffs can make. You might also have worked with excellent PMs you want to emulate or less-skilled PMs whose mistakes you want to avoid.

Customer Focus

Customer Focus is the most important thing to develop when moving from engineering to product management. Engineers and developers can usually pick up most of the other important skills on the job, but a customer focus is one of the defining characteristics of good PMs. This means not just thinking of cool ideas, but relentlessly thinking about the target audience, their hopes and dreams, their needs, and how they're different from you and the other people at your company.

One way to build customer focus is to talk to customers of your current product. Ask the PM or Sales team if they will bring you along on their next customer visit; they usually love to bring an engineer along. When they ask for features or tell you what they need, see if you can dig a few levels deeper to get at the underlying motivation. Basically, you're doing a root cause analysis on the customer's request.

If you can't directly visit a customer, you can sometimes read, or even volunteer to answer, customer support tickets. Working on the customers' side and solving their problems can help you build up customer empathy.

Writing story-like user scenarios for the features you're building is another way to develop customer focus. For these scenarios, put yourself in the customer's shoes and imagine how the feature fits into the rest of their life. It might seem silly, but when you include details about the customer's mindset, you can build products that fit into their lives better. For example, did Sally really turn on her computer today hoping to update her Flash player? Probably not, so maybe the update should happen quietly in the background.

If you're a backend engineer, you probably don't interact with people from different functional groups as much. See if you can treat the engineers who build on top of your work like your customers. Do user studies with them and think about their use cases.

As you're preparing for the product manager role, practice describing features from the customer's point of view by calling out the user-facing benefits. It's important to talk more like a PM than an engineer if you want to be considered seriously for the role.

Think Big

Visionary or strategic thinking is another area to focus on when transitioning to product management. As an engineer, you're probably very focused on what is possible to build. For most of your career, you've needed to lower other people's unrealistic expectations. As a product manager, you need to let go of that instinct and allow yourself to envision a world where you've made the impossible happen.

Teams need product managers who can lead them into the future, building things that have never been built before. At some point in the development life cycle, you'll have a chance to scale back, but you need to start big if you want to build a product that will have an impact.

It might sound crazy, but for any product or feature you're working on, think about how it could change the world. If this is hard for you, here are some tips:

- See if you can tie the benefits to fundamental human needs like safety, friendship, or self-esteem.

- Start your brainstorming with the phrase "If I had a magic wand..."

- Write down your practical objections, then keep going.

- Find a teammate to play the practical pessimist role in your brainstorming.

- Write yourself a reminder to always think big.

- Start your feature planning by writing the press release.

Allow yourself to dream.

Embrace the persuasive elements of communication

Many engineers are comfortable in the world of analytical thinking. As an engineer, it's better to prove things through data than charisma. As a product manager, you need to master both.

We'd love to think that all of our coworkers are perfectly logical creatures, but to accomplish things in the real world, you often need to rally the troops and build up some excitement. A spreadsheet with compelling metrics may not open as many doors as a statement like "I've looked at all of the numbers and I really believe this is the bet we need to make."

Credibility is the currency of the PM role. Sure, each executive and engineer could look at your spreadsheet and form the same conclusions, but they brought a PM on the team so that they wouldn't have to. They want you to do the research and propose a solution. It's your job to cut through the ambiguity to help the team get moving.

The more certain you are of the right outcome, the more persuasively you can speak and the more credibility you're putting on the table. If you end up being right, you gain credibility and can convince the team of bigger things in the future. If you're wrong, they'll be less likely to trust you next time.

This isn't to say that you should swing all the way to the side of charisma. Data is often the fastest way to persuade an engineer, and it can be a much more effective approach before you've built up your credibility. Just don't forget that you have multiple tools and you should use them all to be an effective product manager.

Be prepared for some unexpected ways that the PM role is different than Engineering

When you think about becoming a product manager, there are some changes that you're expecting: you won't be be coding anymore, you'll be responsible for a lot of decisions, you'll spend more time in meetings. There are also some parts of being a PM that are less visible.

Understanding these differences can mean a smoother transition. Before you make the switch, make sure you have considered the downsides of being a PM. You don't want to get a job as a product manager only to learn you preferred being an engineer!

The work is less tangible

In engineering, you get the satisfaction of writing code and seeing it work. As a PM, most of your work doesn't have such concrete output. Stephen, who switched from a developer to PM, shared that "when you're a dev you have all these daily successes: my code works, my build passed. As a PM, you have to remember to look for it: I convinced this person, I got the team onboard, and so on."

You become a focus point for criticism

Ideas can come from anyone on your team. As the PM, you're responsible for crystallizing those ideas into specifics. While everyone might have liked the idea in the abstract, once it's solid you'll start to hear a lot of pushback, no matter how good the design is. As a PM, you need to be able to take that criticism constructively and not personally, and turn it around to make a better product.

You don't have time to do it all

Engineers often get to set their own time estimates and are given time to go deep on a feature. An engineer might spend a whole week working on one part of the code. As a PM, you can't always spend the time you want on all of your work, so you need to prioritize. This means that some of the time you'll have to hand over work that's 30% of what you wanted to do and move on.

Look for openings on your team

One common way for engineers to move into product management is to find a role on their current team. It makes sense; your current team already respects you and has seen what you can do, and you have expertise in the area.

Daniel, who moved from engineering to product management at Google, found a PM opening on his team. "My team had an open PM role that was very data-oriented, and it wasn't clear to anyone how to fill it," he told us. "I knew how to work with data, so I came up as an option. It helped that I had people skills and pre-existing relationships and that I had built up credibility."

Here are some tips for moving into a PM role on your current team:

It's important to let your interests be known.

Find out who would be the hiring manager for the next PM and talk to her. Your team might not currently have room for a PM, but you want to be high on their lists when the position opens up. This also gives you an opportunity to learn what areas she thinks you should focus on improving.

Re-read your prior performance reviews for any potential PM issues.

Have people commented about your being stubborn? About the team never seeming to know what you're working on until it's complete (which could signal communication problems)? About wanting to see you take on more leadership on your features? Take steps to address these concerns immediately and to leverage what PM-related strengths are mentioned.

Start taking on some PM work, even without the title.

Maybe there's a small feature that needs to be spec-ed, or some product decisions you can help make. If your team has PMs, you can offer to help them out. If your team doesn't have any PMs, look around and see what work you think would be useful.

Take on other types of leadership and coordination work.

Stephen, who moved from developer to PM on his team at Microsoft, shared this story: "I was the tech lead on a team with two other devs and ran a cross-team collaboration project for my team. We had a messy dependency on another team that was a big risk, so I took on project management for that. I worked with the partner team, and got the dependency straightened out. That established for my team that I could do project management and people thought of me as a leader. Then when I talked to the PM lead about moving onto her team it felt like a natural fit."

Think about how to clearly mark the change from engineering to product management.

It will be hard to be successful as a PM if you're still handling a lot of engineering responsibilities; you need to pick up some escape velocity. Consider taking time off between the role switch or having some kind of hand off or party to mark the transition. Then you can dive into your new PM role fully.

Find a specialized PM role in your area of expertise

As an engineer, you've probably picked up some domain expertise around the product you're working on. Especially if the industry has lots of specialized knowledge, you can use your experience to boost you into a PM role in the industry.

Perhaps you've been working on security and encryption products. That's a deeply technical area where your experience as a developer will be a natural value-add.

One area where you might have a natural advantage is on developer product or platform products. Since you've worked as a developer, you might have insight into how to build a great product for other developers.

When you're going this route, you want to really highlight how your skills and experience can be an asset to the team. Don't be shy about sharing specific anecdotes and insights about the industry.

Go to Business School

An MBA can help you get a fresh start and round out your engineering background with a business education. This makes you, arguably, the "perfect" PM. Additionally, you'll have a chance to network with up-and-coming business leaders, work on projects from end to end, potentially launch a startup (or just explore startup ideas), become eligible for MBA internships, and be connected with PM recruiters at the top companies. A developer-turned-MBA-student, particularly from a top MBA program, is very well positioned for PM opportunities.

However, an MBA will also cost you lots in tuition money, two years of your life, and even more in lost salary since you'll need to quit your job for a full-time MBA program. This could be easily over $250,000. Given that investment of time and money, an MBA probably isn't worth it if you're solely doing it to get a PM role. There are quicker and cheaper ways to make this transition.

Nevertheless, if you've been pigeonholed into the engineering role, or you have other reasons for seeking an MBA beyond simply wanting to be a PM, this degree can be valuable.

When going the MBA route, pay attention to how business- or marketing-focused you want to be. While all of the big tech companies hire people with MBAs into product management, they vary on how close those PM roles are to the engineers. For example, Microsoft's product manager role lives firmly under marketing (Microsoft's program manager role is more equivalent to the product manager role at other companies.) Ask your recruiter questions to make sure you understand the role.

Transitioning from Designer to Product Manager

Designers have a great background to move into product management. If you feel like you don't have the level of product influence you want, or if you feel like your analytical skills are being underutilized, then you might do well as a PM.

Designers already are familiar with focusing on the customer and designing great products. Some companies have PMs do a lot of the interaction design, so those skills will still get a lot of use.

Practice Prioritization

One of the biggest changes in moving from design to product management is becoming responsible for prioritization. As a PM, you'll be responsible for shipping the product, which means avoiding feature creep and scoping the implementation as you get more information from engineering on the costs.

If you've been relying on your PM to tell you when to rein it in, now's the time to start exercising those skills yourself. Sometimes you need to do things the quick and dirty way, and sometimes you need to cut a feature that would have made the product a lot more usable. As a designer, you can practice by prioritizing the pieces of your designs and discussing them with your PM. See if you'd make the same calls, and if not, try to understand what underlies the differences. Pay attention to what gets de-prioritized, either explicitly or implicitly, because the team runs out of time.

A great way to hone your prioritization sense is to follow up on your designs after they've launched. See if you can talk to customers or read support tickets to learn if your prioritization was right. Are people complaining about a missing feature you wanted to cut? Or are they raving about something you fought to keep in? Most people discover that they can cut a lot more than they thought.

Sharpen Your Analytical Skills

Analytical skills come into product management in two major ways: analyzing what your team should be doing and analyzing how to persuade people to do that thing. As a PM, you need to become comfortable with finding data that convinces people. That data is sometimes from product metrics, sometimes from user research, and sometimes from competitive analysis.

See if you can find ways to demonstrate analytical skills as a designer. For example, maybe you can run a survey to get data that will help influence a design. Or maybe you can learn SQL and start pulling usage metrics. These skills will help you make the jump to PM.

Look for Leadership Opportunities

At some companies, designers are naturally in a leadership role, while at others you'll need to look for places where you can take on more leadership. Can you become involved in roadmap discussions? Are there team meetings that you could offer to run? Maybe you can identify a project the team should be working on, pitch it, and rally people to work on it.

Or perhaps your team's PM is a bit swamped right now. Can you take some of the load off her plate? You might be able to spec something out for her or try your hand at analyzing some data. In many cases, a PM would be happy to help show you the ropes.

Sometimes it feels awkward to take leadership roles when no one has given you permission. In those cases, you might need to "fake it until you make it." Remember that scrappiness is highly valued in PMs. Being scrappy is about being resourceful and finding ways to succeed when the traditional processes aren't going to work. For example, if you find that engineers are reluctant to fix UI bugs, you might come up with a contest that motivates them. That's the kind of story that can really make a good impression.

Transitioning from Customer Support

Paul, currently a senior product manager, used his understanding of customer pain points to move from tech support to product management.

When customers wrote in saying something was broken, he would always ask them what they were trying to do in order to understand the underlying customer need. Then, when he filed bug reports, he included detailed descriptions of the problems, suggestions for potential ways to solve the problems, and ideas for new features that would help. The CEO noticed and asked him to become the company's first product manager.

Paul shared some advice about how to adapt from support to product: "In support you work with one customer at a time to solve their particular need and make them happy. As a PM you have to think about the user base as a whole. You need to think and talk about things in terms of the big picture. Articulate how an issue works in the bigger ecosystem."

Transitioning from Other Roles

Breaking into product management can be hard, but people with all kinds of backgrounds have done it. Sometimes the trick is to be persistent. Focus on how your skills can be applicable to product management, and don't stop reaching out to companies.

Here are some skills you can emphasize when looking for PM roles:

- **Analysis.** Do you work with data at your current job? Are you an Excel ninja? Many software companies are looking for data-driven PMs who can make sense of metrics and draw insights from usage patterns.

- **Customer Focus.** Are you in a customer-facing role? Have you learned how to translate customer feedback into action? Companies love product managers who understand customers and their needs.

- **Business Savvy.** Are you comfortable putting together business cases? Do you know how to size a market? Your experience can be a real asset in making the right business decisions.

- **Marketing.** Do you have a background in marketing? Can you effectively communicate the value of a product? Marketing skills can help a PM design a product that will do well in the marketplace.

- **Industry Expertise.** Do you have deep knowledge of how your industry works? If so, you have a leg up on applying to PM jobs in that industry. Your understanding of the industry means you can be a productive PM in a short amount of time.

- **Helpfulness.** Could your team use some product management help? Do you have some extra capacity to step in? Many people slide into the PM role just by helping out when there was a gap.

When trying to jump to a PM role, compare these PM skills to those required or demonstrated with your current position. You'll need to find ways to patch up gaps in your experience and skill set.

Use Your Network

Don't underestimate the power of networking to help you break into product management. Especially if you've demonstrated relevant skills, having a co-worker speak up for you can make the difference in convincing a team to take a chance on you.

Sara, who went from partner technology manager at Google to product manager at Twitter shared: "it was really critical that I knew people. I got in the door through the people I knew. Originally there wasn't any PM role open; they didn't need any more PMs. These people really advocated for me. They had worked with me at Google and were able to vouch for me. I got into the company as an internal tools PM. Then once I was in the company, I transitioned to a consumer facing team."

What Makes a Good Side Project?

One of the best ways to improve your candidacy for a product management position is to start a side project. The side project gives you a chance to gain experience shipping a product, builds up your resume, shows off your technical skills, shows off your product design skills, and gives you a lot to talk about during your interview. If you hired people to help you, it might also give you a chance to show leadership skills.

A good side project has the following qualities:

- **Fills in the gaps in your experience.** For example, if you don't have a computer science degree, you can show off your technical skills by building a website or mobile app. There are many tutorials online that you can follow to build a simple app for free.

- **Shows off your skills.** If you've got great visual design or savvy business sense, the side project can be a great place to show it off.

- **Is something you can speak passionately about.** Are you solving an important problem? Or did you get to test out some interesting hypotheses? Did you learn a new technology? Make sure you can tell a story about your project.

If you're not technical, you have a few options. You can recruit friends to help you with implementation. You can hire people to build it. You can also make your side project about discovering the feasibility of an idea. For example, you can design the product and create a marketing website that describes it and asks people to sign up to be notified when it's available. You could then iterate on the product idea and see how that affects the sign-up rate.

Another good non-technical option is a design and usability project. Pick a problem in the world or your local community and start talking to people, observing them, and coming up with ideas. Then build a paper prototype and test it out with people - bring it to a coffee shop and ask people if they'll try it out. IDEO, a top design firm, has materials on Human-Centered Design on their website that can help you learn more about running these projects.

When you choose your side project, be aware that interviewers will often ask you a lot of questions about it, so be honest about the motivations for the side project and how far you went with it. It's perfectly reasonable to build a mobile app to brush up on your technical skills, to get experience launching a product, or to try out some new design paradigms. In fact, those are great reasons; they show a passion for learning and experimenting.

Be prepared to talk about how you would improve the project if you were to continue working on it, why you made the choices that you did, and how it was and wasn't successful. If you have any positive metrics (user signups, revenue per user, etc.), these are good to discuss.

Your side project should be listed on your resume as well as on your website, if you have one. It's great if your side projects were successful, but just having done something shows a lot. Don't be afraid to list "unsuccessful" side projects.

Career Advancement
Chapter 5

Congratulations! You've landed your dream job as a product manager working with a team you love on a product you're passionate about. What now?

Tips and Tricks for Career Advancement

Working hard will help you be more successful, but it's not enough. The person with the most hours is not always the one who lands the promotion.

What's equally or more important is how you work and what you work on. As you build your career, these tips will help you achieve greater success.

Ship great products

As a PM, the biggest measure of your success will be the products you launch. More so than any of the mocks you drew or specs you wrote or bugs you triaged, you'll be recognized for how well your products do in the market. If customers love your product and you've got a viral growth rate, you're in good shape.

At the end of the day, people won't know all of the big and little things you did to make the launch possible. They will remember that you led the team that delivered a successful product.

Get some launches under your belt

The product life cycle, from planning through implementation to launch and beyond, sets the main rhythm for a PM. Each part of the life cycle requires different product management skills, so if you want to learn and improve, you'll want to go through the whole life cycle several times.

The product life cycle varies in length from team to team. When you're starting out, you can pick up experience more quickly by finding teams with shorter launch cycles.

Become the expert

When you join a new team and all of your teammates seem so smart, it can be tempting to just shuttle questions to other people since they know so much. But if you're just passing questions to other people, you're not really adding a lot of value.

Instead, make sure you really take the time to become the expert on your areas and your customers. Think about what kinds of research you can do to really understand the space. Maybe you'll want to visit customers or put together a competitive analysis. You can look at data and metrics, or you can chat with the sales team. Talk to as many stakeholders as you can, and learn the background from your team.

Once you've done your homework, you'll feel like you really understand the space and you can confidently make decisions.

Find teams where you can pick up new skills

"Seniority is all about experience, but there's a catch," says Chrix Finne, a senior product manager at Optimizely (and formerly Google). "You can control how fast you accumulate experiences."

If you've mostly worked on improving a mature product, consider joining a team building a new, unlaunched product. If you've always worked on consumer-facing products, consider trying something business facing. Look at your skillset, decide which skills you need, and then find a place where you can learn those skills.

Pick the company where you'll learn the most

At different stages in your career there will be different things you need to learn. You can optimize your learning by choosing a company that's set up to support you.

When you're a brand new PM, you'll need to learn the basics of product management. You might want to pick a company that has several strong PMs to learn from.

Working closely with PMs is an excellent way to pick up the trade. After you've been a product manager for a while you might want to increase your responsibilities and learn how to work more independently, so you might want to pick a smaller company.

Choose a growing company

At a growing company, new opportunities are always opening up, and you quickly become one of the more senior employees. This means that even if you had to join a different team or take on different title from what you wanted, you'd likely get a chance to transition soon.

"If you're in a growing company and working on a product you're excited about, there will always be more opportunities." said Sara, a PM at Twitter who started on internal tools and then moved to customer-facing teams.

Find a manager who believes in you

Many great PMs credit their success to great managers who gave them opportunities to prove themselves. When you're choosing a team, don't just look at the product - also consider who your manager would be.

You can often talk to other PMs at the company to learn who the best managers are and who are the managers to avoid. Once you have a good manager, show them you are reliable and can do good work. Then, talk to them about how you want your career to grow, and be brave enough to take on the challenges they give you.

Focus on your own efficiency.

As a PM, you're bombarded with tasks, and you need to know what you can drop. You want to be responsive to your team and never become a bottleneck, so it's important to prioritize how you spend your time.

If you have trouble getting to "Inbox Zero" (no messages in your inbox), try learning a time management system like Getting Things Done. A few small changes to your routine can sometimes make a big difference in how organized you are. As you become more efficient, you will feel like you are gaining hours in the day.

Understand how your role fits into the company.

Career advancement as a PM usually means expanding the scope of the area you work on. You might start by working on a feature, move to PM-ing a larger area, and then eventually own a whole product or even a product suite.

To grow in this way, you need to understand how the pieces fit together to see the bigger picture. Ask "How does my feature fit into the product?" and then, "How does this product fit into the suite of products?" Think about the connections to get a broader picture.

Help your team with something tangible early on.

Most teams are a little bit suspicious of new PMs. They're worried you'll create busy work for them or slow them down in other ways. It doesn't help that most of the work a PM does is behind the scenes, so your coworkers might not see how hard you work. You can counteract their fear by focusing on helping out your team when you're new.

Look around your team to find some grunt work you can take off of someone's hands, or do some research that people have been putting off (but really wish they had done). This is an easy way to get off to a good start with your new team and earn goodwill that will help you in the future.

Work on something that's important to your team and the company.

Many companies will start new people off on unsexy work because it's a low-risk way to see what you can do. In these cases, you want to do an amazing job on your unsexy piece and prove you can be trusted with more important work. Then you can find areas you think are important and offer to contribute to those.

Make sure you draw the distinction between teams that are just unglamorous and teams that are truly unimportant. Infrastructure teams often don't sound like a fun place to be a PM, but they're critically important to the company. The improvements you make as an infrastructure PM can be magnified throughout the company, so they can be a great place for career advancement.

Take on cross-team or company-wide tasks.

At some point in your career, your visibility across the company is going to matter if you want to be promoted to higher positions. Sometimes you can get this visibility just by launching big projects that are important to the company, but there are other ways to get your name out there.

By leading and doing a good job on big company-wide projects such as UI reviews or goal setting, you help more people across the company think about you as a good PM. Similarly, you can teach a class or present at an all-hands. When a committee is deciding if you'll get promoted to the next level, it's great for all of them to have heard your name and appreciated your work.

Another reason this cross-team work is valuable is that it helps you form relationships with people throughout the company. Sometimes work between teams can be tough, but it goes much more smoothly if you already have a friendly relationship with the PM on the other team.

Define and measure success.

One way to really stand out as a PM is to get more concrete about what success means for your team. Depending on the project, success might mean more user growth, increased revenue, or increased customer engagement. For other projects you might have a more specific feature-based goal.

Think about what you're aiming for, communicate that to your team, and measure whether you're hitting it. This makes it clear when you're achieving your goals and helps you learn when you don't. Once you're clear about your goals, you can better prioritize what work is helping you and what work is unimportant.

Don't let your team do unimportant work.

Sometimes big companies lose track of all the projects going on, and you might be assigned as the only PM for a team that's really not delivering much value for the company.

Ideally, you'd redirect the team to work on something more important. If that's not possible, consider suggesting to your manager that the team be shut down (make sure you do your research and can support this) or consider switching to another team. It will be hard to get promoted launching a product that doesn't matter.

Don't just do what's asked of you. Get the job done.

As a new PM, it can be tempting to think of your work in terms of deliverables such as proposals, specs, and analyses, and then to think you're done once the document is written.

Those documents aren't your job; they're just the tools you use to get results. Make sure your engineers use the specs to build features, or rewrite the specs or find another way to get your ideas across. If you made a proposal, did you convince your team to pick it up? If not, find another way to make your point.

Demonstrate you can consistently deliver work at the next level.

Most of the larger companies have an explicit career ladder with descriptions of the skills needed at each level. These ladders can be frustrating because they look so explicit and concrete, and yet PMs can meet all the requirements for their current level without being promoted.

Two things are going on here. First, you need to meet all the requirements for the next level before being promoted into it. Second, you need to have earned a reputation for consistently and repeatedly delivering work at that level. Sometimes you'll have shown you can deliver on a requirement once, but your team doesn't have the confidence yet that you'll be able to repeat that.

Find a mentor (or mentors).

PMs have all kinds of different skills. Some people are really good at coming up with a vision for their team, while others are great at data analysis or design. Identify people who you think are really strong in an area and reach out to them.

When reaching out to a mentor, be specific. Maybe you're trying to build a case for your idea and reach out to someone who's really persuasive to ask for help on how to frame your proposal. Maybe you want to show your mocks to someone who's really strong in usability. Being purposeful makes the relationship better for both of you.

Build credibility.

"Credibility is the currency of a PM," says Daniel, a Google engineer who became a Google PM. As a PM, you want to gain the respect of everyone you work with, whether they be other product managers or people from other roles. You especially want to earn a good reputation with leaders of other teams so they'll want their people to work with you.

The most straightforward way to build credibility is delivering results. Your teammates all want a good outcome, so they'll naturally start out second guessing your opinions, asking lots of questions, and suggesting different ways of doing things. However, over time they'll start to see that you're showing good judgment and getting things done, and they'll feel comfortable trusting you.

Another way to build credibility is paying attention to people's perceptions of you and ensuring that you're creating the perception you want. Make sure you're building a reputation as a smart, skillful, competent, and dependable person with good judgment. It can be hard to get feedback directly from people on your reputation, so it might be worthwhile to ask your manager.

Q & A: Fernando Delgado, Sr. Director, Product Management at Yahoo

How did you make such a big jump from being a product manager at Google to being a senior director at Yahoo?

I decided to move to Yahoo relatively early after Marissa joined as CEO. I was taking a big of leap of faith. I think joining so early, when it was risky, helped me personally figure out a role I was excited about and gave me a bit of leverage.

In large companies, it's really hard to make big jumps if you stay in the same company. They have really set ways about how they do promotions. There are people who are put on the fast track, but they still have to go step by step. If companies didn't do it that way, there'd be lots of upset people.

It's pretty obvious to me that, in big companies, the only way to make big jumps is to jump to another one. You get a big reward if you take a lot of risk. You can go to a startup. Or you find a very unique opportunity like I had, where a large company gets a new CEO and is willing to look at things with a new lens.

This might cause you to conclude that the only way to make big jumps is switching every year or two, but that's not what I'd advocate. You want to become an expert in certain areas and be given the opportunity to get these chances. If a company sees that someone's never been with the same company for more than 18 months, you'll assume it will be the same for them and they won't want to make an investment. So stay a good few years before you make those jumps.

What were some of the key breakthrough moments in your career?

When I went to work on Google Maps in Zurich. That was a breakthrough moment for me. In my first year as an APM, I worked on Google web-search quality. When I moved to Maps, one of my first assignments was search quality on Maps.

There were some ideas and frameworks that were working well on web search, but they weren't being used much in Maps yet. That meant that I could bring in some of the lessons from web search to Maps—some of the things that we knew worked.

I was able to work in a product that's the most successful in its industry [Google Search] and take lessons I learned there to another product. Because I had first-hand experience with the ways a cutting-edge team did things, I had value to bring to a different product. I brought a new rigor and emphasis on techniques that ended up being successful in a new product. Some things were different of course, but at the end of the day, the fundamental ideas were the same and they worked well.

Another part of this was that I moved to Zurich. A lot of people in the Zurich office hadn't had a chance to experience how web search did things in person. I brought a new perspective that most people hadn't encountered. This allowed me to have credibility with the engineers there.

At some point, you become well versed. You have exposure and see the patterns, and this helps you do things better and better every time. It gives you a leg up in your career. In Zurich, I had the self-realization that even though the product roles are different, there's value in bringing over lessons from other places, especially if you've had an opportunity to work on a leading team.

Another example of this: Facebook and Twitter have been really successful at having and nurturing growth teams. Some of that culture has started to have an impact on and show the way for startups. PMs who have worked on product growth have skills that would be applicable. Even though growing from zero users to five million users is different than growing from 10 million users to 50 million users, a lot of the general lessons can be applied.

What advice would you give to a PM who wants to advance in their career?

It's important to realize that, in a lot of PM roles, you get to a tipping point, after which making decisions and working with a team become a lot easier. Things flow more naturally. You get less cynicism and skepticism from engineers, and it enables a friendlier environment.

Reaching that tipping point should be a big thought when you start a new role. Circumstances can make it easier or harder. Currently, I'm leading mobile development for the Yahoo mobile app for iPhone and Android. The team started as a designer and myself and has grown significantly. I assembled my team and that got me past the tipping point and made a lot of things easier.

I've been in projects where I felt like I could not get past that tipping point. The decision at that point was to move on to a new project. On the Android Market team, there were a lot of difficulties in terms of strong opinionated leadership, but no single clear leader. The consequence of having such team structure was making compromised decisions, often picking the wrong tradeoffs. Individual engineers could see this happening, which made it difficult for me as a PM to get past the tipping point.

Larry, the CEO, tried to get teams to have a single decision maker at the top, but on my team that figure wasn't well identified. A lot of the engineering leadership had strong product opinions, which can be fine, but I'm also strongly opinionated. Working on the team involved clashes that were hard to resolve without conflict.

Android was a very successful product, and people had a culture of working really hard. I felt like because of the structure of the team, not everyone could get rewarded for their work. This coincided with my first daughter being born. I was spending a lot of time at work in an intense environment and was not getting rewarded for it.

I saw I wasn't going to go past the tipping point because other people had been with the team longer than I had. When push came to shove, they got their way. I was coming home from work frustrated. When I joined Android we had 50 million users and we grew to 350 million users while I was there. This was something to be really proud of—to contribute to this great, successful product, which was growing exponentially. But I was sacrificing my work/personal balance with no clear path. So I decided to switch teams.

I've been more conscious about figuring out the tipping point and how to pass it. Some factors are in your control, and some are out of your control. Pick a place where you can stay long enough that you've been on the team longer than most people. Think about how long you're willing to stick with a product. You cannot speed up time, but you can choose a place where you're more likely to become a senior member of the team.

As an early tip, be as much of an observer as you can. Don't try to disrupt or change things too much early on. You'll come to a product and have your conceptions of what needs to change, but most times there's a culture of making decisions and reasons for why things haven't been high priority in the past. You don't want to pull the trigger too quickly.

Understand the context of things. Be more inquisitive. Instead of telling people what to do or trying to make decisions, try to ask questions. Why is this the way it is? Try to understand the context and history instead of being the new dictator of the team.

Another tip to get that tipping point early on is to demonstrate value to people. Make it so the fact that you're there is driving something that wouldn't happen if you weren't there. There's often administrative or boring work that isn't getting done. People agree that it should get done and has value; it just isn't at the top of anyone's list.

For example, let's say you're working on a news product and you need to show the logo of the news sources. Currently it's a manual process to pick up the logo, so offer to do this for them. Find a scalable way to do it, maybe scraping, maybe using Amazon Mechanical Turk. Remove busy administrative or boring work from people, but make sure it's in a way that actually adds value to the product.

Another tip for more senior PMs: Figure out your own framework and principles for how you make decisions, and communicate that as often as you can. You start developing these frameworks, but sometimes they're not obvious to people around you.

Make it clear to the people around you why you're making a particular decision so they see that you're consistent with your decisions. There's nothing engineers hate more than subjective decisions that change from one day to another. If you develop that framework and those principles, it helps people realize that you are consistent.

As an example, Adam Cahan, the Senior VP of the mobile org, has an incredible eye for animations and transitions on mobile. Every time you bring him a new build of the app, he might notice that one of the transitions is inconsistent. He doesn't have a strong opinion on how you do transitions, but he cares that, once you pick a metaphor or general mental model of the app, you stick with it.

If you pick transitions where you get a sense of depth and go deeper and deeper, that's fine, but then never violate that principle when you open a settings pane. Take the metaphor for your app and use it across the board.

When you come to him with a fancy animation for something and he tells you not to do it, you might feel resistant at first. But because he's communicated his principles so clearly, you realize that he has a good reason.

Another example is that I try to make sure that, as much as possible, when you open the mobile app, you get straight to the value with as few bumps as possible. So if we're adding a feature that won't be used by everyone, we avoid welcome screens and tutorials. My position is always: don't show it up front. You can educate people, but don't put the bump up front.

Over time we realized that if you stick to that principle, you find other ways of doing things. So, we have shortcuts for power users. But we'll only show in-app hints and popups if we know for sure that you're interested in the feature. We wait until you used a feature in the non-shortcut way and used it enough that our popup hint about the shortcut will make sense to you.

Q & A: Ashley Carroll, Senior Director of Product Management, DocuSign

Tell me a little about your career path.

I started at JPMorgan out of undergrad, because that's what you did at that time with an econ degree. I was in the San Francisco office so I ended up sitting within the Tech, Media, and Telecom group. It was great because I learned about all these great internet and software companies.

Following the Shutterfly IPO, which was led by JPMorgan, I joined Shutterfly as a business analyst. It was a great first role at an operating company because I supported a variety of functional areas with reporting and analysis, and I got to see how they spent their time. I got really interested in seeing how new features and changes to the UX (user experience) would affect customer engagement metrics and ultimately revenue. I also found that I really enjoyed working with engineers and designers; I wanted to be more involved in creating things to delight customers. A couple of years later, I left to pursue an MBA, mostly because I felt it would be an efficient way to move from analytics to product.

I did a couple internships before and during business school - at oDesk and Amazon Web Services - which were a great way to get a taste for different company sizes, cultures, and sub-sectors. After business school, I joined Survey-Monkey's product team. I started out as an individual contributor focusing on growth initiatives, like launching new plans and overhauling new user onboarding. I eventually grew into managing the platform team, which was responsible for things like identity, billing, and enterprise features.

I left SurveyMonkey a few years later to lead product at Optimizely where I built the beginnings of the product team from the ground up. The role turned out to be a bit different from what I expected and wanted so I left after about six months. I decided to do some consulting work to get to know any potential team (and manager) before committing. After a couple months of spending two days per week at DocuSign, I was hooked. I converted to full time and am currently overseeing growth, which is really exciting given the stage of the company and viral nature of the product.

What were some of the key breakthrough moments in your career?

When I was working as a business analyst at Shutterfly, a PM left. Because I was already familiar with the metrics for that product line, I had the opportunity to step in and gain some PM experience. Having the opportunity to get some product work while at Shutterfly was crucial. Post-business school, I was able to reference this experience and provide concrete examples around product launches and tough UX decisions.

I can't say enough good things about SurveyMonkey. I feel so lucky to have had the opportunity to work there when I did. The leadership team is really great about growing talent and giving people the room and support they need. I would get a new project or responsibility, have no idea how to do it, figure it out with help from my boss and the team, and then take on additional new responsibilities before I had time to get bored.

My current role was very much the result of being at the right place at the right time. I was introduced to DocuSign just as they were building out their product team. The company had a strong lineup of enterprise software veterans, but they were in the market for product folks from more consumer and UX-focused backgrounds. The result is that my role is a perfect mix of things I've done before that I'm very comfortable with and new challenges that push me outside of my comfort zone.

What advice would you give to a PM who wants to advance in their career?

Join a company that's experiencing (or, even better, about to experience) hyper-growth. The business will be growing faster than the team can scale so you'll have lots of room to grow in terms of responsibility. The venture capital community can be a great resource here; they value meeting talent (you!) just as much as you value getting job leads so it's a great symbiotic relationship.

Make who you work with your top priority. They're your best resource for learning. And since things in tech can get stressful, it's important to make sure they're genuinely good people. The best way to vet people is to work with people you've worked with before. That's obviously tough in the beginning of your career, but with enough scrappiness you can probably find second or third degree connections. Another option is to "try before you buy" via consulting.

It's important to have a manager and executive leadership team you trust and believe in. It's tempting to chase more senior roles at a startup early in your career, but I'm really thankful for the time I've spent working under experienced exes. At this point of my career, I'd rather work for someone I can learn from and ask questions of than be at the top.

It sounds trite, but it's a small Valley (and beyond). Good employers will do official and unofficial reference checks, so it's important to actually be the person you want to be. That doesn't just mean working hard and acting smart; it's more being honest, staying humble, and genuinely caring about the team. Those are the things that make me want to work with someone.

Q & A: Brandon Bray, Principal Group Program Manager, Microsoft

Tell me a little bit about your career.

I've always been in program management. I started in Microsoft Office as an intern on Outlook over two summers. After college, I came to Microsoft on Visual Studio on the team where C# began, which also happens to be the team where C++ was.

I started on the C++ backend, so it was very technical. Some people think of program management as just frontend, but I worked with AMD and Intel, all of the silicon vendors. It was really fun.

After doing that for a while, I switched to the other side of the C++ compiler. I designed the language for the C++ CLI [common language infrastructure]. I got to travel all over the world and work with ECMA and ISO and all the companies that are moving the language forward.

Then I decided to broaden my experience. I took on a project management role and became a lead at the same time. I took the project management role to deliver Visual Studio 2008. This was still a program management role, but specifically around release management. Program managers usually do just a little bit of project management and spend a lot of time on design. Release managers pay attention to exit criteria, and all of the other pieces of project management to make the entire release go well.

After that, I went to China for a year and built up a PM team in Shanghai. That was when I realized that most people learn to be PMs from watching other PMs. In Shanghai, there weren't many other PMs around. So I had to find another way to help these PMs learn.

Next, I came back to the US and became a group program manager on the .NET Framework. Group program manager is the second-level manager. I helped ship Windows 8 and then moved to the C# and Visual Basic team. The team had a new compiler, but it didn't look like it was going to be ready on time, so I helped get it over the shipping hump.

And most recently I switched to a new, unreleased hardware project. I decided to switch back to an individual contributor role for a change, and I'll be working to make sure we have a great development platform.

What were some of the key breakthrough moments in your career?

One that definitely was a highlight of my career was writing the language specification that eventually became a standard for C++ and the CLI. I had to work with people from different companies.

This was where I learned that whoever writes things down has the power. People in different disciplines such as test and dev think that PMs make all the decisions. Then when they switch over and they wonder how they got to make decisions, and it's because whoever writes things down records history. The ways you write things down and what you write down define history. The power of the pen.

I got to have more influence on the way things should work out. Also, writing is thinking. If you go through the details of writing things down, you end up thinking about the corner cases.

Another moment that I talk about is around the impact you can have really early in your career. One of the first things I did on the C++ compiler was the /GS feature [buffer security check]. This feature was built in response to the buffer overrun attacks. The team had built something to counteract buffer overruns, but it missed a lot of cases. I looked to see what was needed to make it really useful.

Then, I talked to all of the teams at Microsoft and convinced them to use the new compiler and rebuild their code with this switch. It was a lot of work, but it made a big difference in the security of our products.

The last key breakthrough has been as a manager in the past few years. Shaping how PMs think and shaping the culture of a PM team have been fun for me. Working on some of the really technical teams it's really easy to forget that there are different styles of customers out there. There are developers, end users, IT staff, and so forth. We had to really focus and remember that electrons are not our customers; people are our customers. I've helped my team remember to think about how people experience what we do.

Were there any things you did that helped you get promoted?

My time as an intern helped me go from PM 1 to PM 2 quickly. I'd learned how to be a program manager from my internships. I was promoted into the senior band around the same time as I became a lead and a project manager. I'd written down that I wanted to be a lead at some point, and the team came to me and asked me to be a lead since they needed someone to do the work.

For being promoted into the principal rank, I took on a division-wide procedure - what we called Fundamentals and Tenets. It was a system to go through all the pieces for a successful product: security, performance, reliability, world readiness, privacy, compliance, things like that. So it was a process to make sure all those things happened, but it wasn't going well at all. They gave it to me to run.

I worked at the divisional level with all of the product teams. I had to engage with 300 people to get things moving along, and I got it to the place where I turned the problem around. It was fighting a fire. Since that project had a really large scope, that took me to the principal level.

What advice would you give to someone interviewing to become a PM?

The thing I tell everyone is that a PM is an expert in their customer. That's what sets them apart from the developers and testers. A PM on Excel has to be the expert in the kind of things people do on Excel. Those people might be quants or other kinds of number crunchers. Developers on Excel don't need to understand quants, but the PM does.

In my space, those customers are developers. And that brings me to another point: you have to really like your customers. If developers are your customers, you need to be able to get into their mindset and be the customer yourself. If you're a PM for a game, you should really sympathize with and understand gamers. Same thing if your customers are database admins.

When I'm interviewing, the thing I check for, no matter what, is passion. Do you care enough about your work that you're spending your free time learning about it?

As advice for someone new, make sure you find a job where you're interested in doing something outside of work. Whether it's going home and working on a project or a going to a meetup or going to conferences. Something that extends what you have to do on a day-to-day basis. If you have that passion and a drive for learning, you're set up to be an awesome PM.

Sometimes people ask me if you can work on a team when you don't already know the area. If you're interested in the area and have a drive for learning, you can be successful in any group. People learn really well when they have an objective in front of them. Great PMs have an enormous impact on their team's efficiency, so the people on the team are incentivised to help the PM learn.

As another interviewing tip, when I'm interviewing and I want to tell if the person should be a dev or a PM, I'll usually ask questions that can go in two different directions. You can either start designing the solution and building algorithms, or you can step back and talk about who the customers are, what the goals are, and start defining success.

Good PMs can start down the implementation path, but pretty quickly they realize they need to step back and talk about the customer. If they don't, it's pretty obvious they don't have the inherent starting ability to be a PM. It has to be a trained skill.

If you're interviewing to be a PM, it's good to look at every problem starting with "Who is the customer?" and "What is success?" I do that all the time. I'll be at a stoplight in a traffic intersection and will think "How can I make this better? Who am I making this better for?" These problems show up all the time, so you can train yourself to think this way.

Q & A: Thomas Arend, International Product Lead, Airbnb

Tell me a little about your career path.

I studied math and computer science in Berlin. My first job was with IBM. Originally I was an engineer. At the time, there weren't any product managers.

I found that what I liked most in my job was to go out and explain the product in non-technical terms to our customers, and then come back and explain who our customers are to the rest of the team. I'd talk to users and potential users and try to find out what they wanted. Usually the first stage was just listening to them complain.

It took me a long time to figure out what product management was. I was happy with what I did, but from my manager's point of view, I was an under-performing engineer. At my performance reviews, he'd say, "In terms of engineering you aren't one of the top performers. But whatever you're doing, keep doing it." So I kept doing it, and it was a big relief for me when I heard about the product manager role, since I realized that was what I was!

I was a PM for SAP for several years, and then switched to the business side of things for a bit. I worked with the CEO on company strategy for a few years. That was a great learning experience, but I missed building products.

So I went back to product and design. After 11 years, I had learned so much and the company had grown so much that my interests shifted toward the web and to much smaller companies.

I left for Google and worked there for five years. I worked on internationalization and gave them the 40 languages initiative, and then worked on iGoogle on Marissa Mayer's team. When I left, I left on a high note.

I joined Mozilla for a year. I was fascinated by Mozilla and nonprofits. It turns out I loved the idea but not the execution, so then I moved on to Twitter, where I stayed for one and a half years and worked on several products.

I've been at Airbnb for 7 months. I'm the head of International. My goal is to make us successful in 192 countries. The role includes not only localization and translation, but also includes the strategy in different countries. I get to not just build one product, but to work with the other PMs to build the best product in all countries.

What were some of the key breakthrough moments in your career?

There were many moments that helped me become what I am today. One of the big things was talking to a lot of awesome role model product managers. When I joined Google, I talked to Sundar Pichai and a lot of people who had been there for a while and whom I really respected for being awesome product managers. Not to copy their style, but to learn what worked and didn't work. I learned so much from them.

Now I'm on the other side of those requests, and I meet with everyone who contacts me. I strongly recommend to everyone who wants to become a PM that they build a network. You learn to write by reading. Look at good products, see how they build products users love, then contact the people who work there and ask them how they did it.

Another big moment was when I worked on design services, such as an internal IDEO at SAP. The team was awesome. I participated and ran workshops with IDEO. It just blew my mind; in only one day of a workshop, I learned empathy with your user and rapid prototyping. They open your eyes for unmet user desires and teach you how to observe without jumping to conclusions.

You get thrown into a new problem like redesign a bookstore. With little guidance, you go out and talk to users, take pictures, come back and synthesize, prototype, and bring your prototype back to user to see how it works.

Young PMs can do that on their own. If you live in Mountain View, take a potential task, such as how to make the Caltrain ticket machine a delightful experience, and go through those steps. Observe, take pictures, talk to people using the machine, and find out who they are.

Figure out your personas - maybe they're commuters or tourists. Then synthesize the problems. Maybe there's glare on the display. You synthesize, prioritize, simplify down to themes, and then build prototypes. You can even make paper or cardboard prototypes. Then go back to those users and ask people if they want to try your prototype. You can do this alone. It's better with other people, but you can do it yourself. One guy I know started a blog and did this with lots of things; it was addictive for him.

Now I'm on advisory boards for startups. I try to teach them to understand who their users are and why would those users would care about the product. I see so many pretty solutions in search of a problem. I talked to a company who wanted product advice. I asked them who their users were, and they said wedding planners, bakers, personal trainers, the flower shop around the corner. So then I asked how many of those customers they'd actually talked to and they said zero. So I told them the first thing they needed to do was go out and talk to their users and find out who their users are.

What advice would you give to a PM who wants to advance in their career?

Know what you're passionate for and passionate about.

There are some pretty good tools you can use to learn who your users are and create hypotheses about their desires and come up with ideas about what the solutions might be. There's a methodology of storyboarding and personas coming from user research that you can learn.

It helps to have a CS degree or be technical. You need to work with engineers and earn their trust. If they have to dumb it down for you, you're lost.

If I develop a new line of cars, I don't have to be an engineer to build the car, but it would help if in my spare time I can at least change a tire and know what's going on under the hood, and ideally I can name all the parts and get my hands dirty to change some of them.

It's important to know what's under the hood and have an interest in it. If someone doesn't have any interest in the technical side, then maybe the technology field isn't for them. How would you feel about teaching yourself some Java by buying a book, installing eclipse, and building a simple mobile app? If that would excite you, that's great. If you think, "Do I really have to?" you're probably in the wrong industry.

It's good to know as much as possible about design and user research. A good PM doesn't only create delightful experiences but knows how to measure success and define success. User-focused, metrics-supported decisions are always the best.

Q & A: Johanna Wright, VP at Google

Tell me a little about your career path

I was a math major and then got my first job as a programmer at a small software company supporting the financial services industry. While I was there each of the new grads had one on one meetings with the company execs. When I met the Product Manager, Tom, I was blown away that he actually knew what the company should build. I dreamed that one day I would have the skills to know what we should build. That's when I decided I wanted to become a Product Manager.

I tried to become a PM in all kinds of backwards way—first I became a QA [Quality Assurance] manager, then I became an engineer again. Then, another startup doing an Internet platform was trying to recruit me. They wanted me to join as a QA manager, but I said I'd only come as a product manager or a programmer. And that's how I became a PM.

When the internet bubble collapsed that company went out of business and I moved to California. My husband and I decided to bicycle from Brooklyn to Los Angeles. I didn't think I'd be able to get another PM job when the economy was such a disaster, but I thought I maybe I could get a project management contract. I was lucky enough to meet a friend of a friend while hiking and she hired me. Six months later I started business school at UCLA.

For me, business school lent a lot of credibility. My startup was pretty unknown, my undergraduate school, Barnard, wasn't well known for CS, and my networks weren't strong. Business school wouldn't be right for everyone—people who went to Stanford, or had prestigious jobs at other companies would not have as much of a credibility gap and may not need business school as much. But for me, given the job market at the time, it made sense.

After business school I went to Google as a product manager. I had to hustle to get the job. There was a while where I had a weekly recurring entry in my calendar to "call Hilary", a woman from Barnard's alumni database who worked at Google. Eventually, once I got some other job offers, Hilary was comfortable enough to pass me along to a recruiter.

At Google I've worked on a lot of teams. I started on Ads UI and then moved to the Search team where I launched Universal Search, Google Instant, and Knowledge Graph. After that I was ready for a new challenge, so I moved to Android where I'm now VP of Search and Assist

What were some of the key breakthrough moments in your career?

My first PM job was a real breakthrough. I worked at a small company and could really define what a product manager was since I was the only one. The way I got that job was interesting. Back when I was a QA manager I had brunch with a couple of the founders of the company. I was really excited about my job and had a whole philosophy about how teams should be set up and run. Just from that brunch and seeing how passionate I was they wanted to hire me.

The defining career moment for me at Google was working on Universal Search and launching it. It was an important project for the company and was known as a project that people hadn't succeeded at before. Once I launched Universal Search I was promoted to Director.

Being a QA manager was a grounding experience in my early career. It gave me perspective on how to use a product, know your users, have empathy for customers. My experience as a programmer was also really helpful to understand what programmers go through. Working anywhere in the software development process is a big help to being a PM.

What advice do you have for people who want to become PMs and advance in their careers?

Do something fun. Do something important. I don't think there is a single answer to career advancement. I have always tried to find jobs that will be fun and keep my brain engaged all day long. The biggest career leverage I have found is when I do something important for my company.

Even more advice.

Communicate what you are doing. In a big company you really need people above you who will allocate resources and get conflicting projects to be on your side, so messaging what you are achieving becomes important. With Universal Search, a baby step was the very first user study. We had to have a demo ready for the user study, and that was really important for the effort. People could see the product working and this helped communicate what we were trying to do.

Don't give up. When job hunting it's important to be willing to talk to lots of people and not worry about rejection. It wasn't that easy to get my job at Google. I was the last person in my MBA class to get an internship. Take heart, all these things can turn around.

Find a boss who believes in you. At Google It was really important to me to have a manager who believed in me. It was great to have Marissa Mayer as a manager - she took chances on me. She gave me a chance to manage people when I hadn't before. When I was 9 months pregnant, she suggested I present our search strategy to our executives. This was really important for my career because it had me in peoples' minds as the person planning the search strategy. That's all thanks to Marissa.

What does it take to be a great PM?

The great PMs I've worked with are very different from each other. That's why it is an awesome. You see people who are full of energy and able to motivate people through their ideas, people rooted in technical capabilities, or people people who get things done through relationships.

That said, all great pms are goal oriented. They're able to get things done, focus, prioritize. One mistake junior PMs at Google (and probably at other other big companies too) make is to go to too many meetings and think their work is done. A good PM defines what it takes to achieve her goals. Less effective pms let their schedules sway how they get things done and it gets in their way.

Great PMs care about their users. They use their product. They do not finish when the spec is done but make sure to to use a working version inside and out. They have great follow through to get things done.

Q & A: Lisa Kostova Ogata, VP of Product at Bright. com

Tell me a little about your career path.

I came to product management from a somewhat unorthodox background. I earned a dual degree in international relations and finance from the University of Pennsylvania / Wharton and upon graduation went to work on Wall Street for a few investment management firms.

I spent my first five years in private equity and venture capital investing, developing a skillset that in retrospect was very useful in my product management career: evaluating products and teams, figuring out which ones to invest time and resources in, and calculating return on investment. It was great training for a critical mindset and for having to constantly think about tradeoffs and opportunity costs. It was also a good practice ground for communicating hypotheses clearly and succinctly to different audiences.

In 2007, I went to Harvard Business School (HBS), where I evolved my goals for what I wanted to do next with my career. I had always known that while I found investing very intellectually stimulating, I was really itching to be part of the creation of products and services—to be hands on.

I also got engaged during the first year of business school. Since my fiancé was deeply rooted in the Bay Area, I knew I would be coming back to San Francisco to pursue my career. The explosion of internet companies such as Facebook, Twitter and LinkedIn really captivated me. I spent a lot of time studying them, working with some of them on field projects and in general, getting excited about the space.

Between my first and second year at HBS, I interned at Google. Now, Google is one of those companies that have strict educational requirements for their product manager role and they require a CS degree. However, I was fortunate to land an internship in their online sales and operations group. The position was more closely related to a project manager role but I was grateful for it; it gave me great exposure to the company and the space, and it got my foot in the door working in operations in tech.

Due to the economic collapse at the end of the summer in 2008, none of the MBA interns at Google got offers. This was a blessing in disguise for a lot of us. It allowed us to be more nimble and creative when pursing career opportunities. Some of my classmates started their own companies; others graduated without an offer.

I was part of the latter group. I figured I'd move out to the Bay Area and pound the pavement looking for opportunities with startups and growing companies that didn't recruit on MBA campuses and needed immediate help.

Looking for a job was a roller coaster since I didn't fit the traditional background of a lot of product management roles at established companies – I had worked in tech for only a few months and people in large organizations were steering me towards the product marketing manager role.

After a while, a few of my contacts pointed me towards Zynga. At the time, the company was exploding: Farmville had just launched, and the CEO, Mark Pincus, was a big believer in developing product managers from business backgrounds. The company needed immediate help and they were hiring on potential, not experience. I jumped at the opportunity. After two days full of grueling interviews, I was in.

What followed were a few months of late nights and "keep your head down" type of work. It was the toughest boot-camp experience imaginable and I saw a lot of product managers quit within weeks. But I knew I had to stick it out to learn the ropes, and I did.

I spent almost four years at Zynga, which was packed with the equivalent of seven to eight years of experience. People used to refer to a year at Zynga as a "dog year" since we were shipping new games, features and products at a record pace I haven't seen anywhere else.

Towards the end of my time at Zynga, I was running a team of a dozen of people and we had built some of Zynga's most important cross-game channels and products on Facebook. I loved my team and had developed a wonderful relationship with them. Some of them were even teaching me coding on the side.

As much as I enjoyed my team, I knew that time was running out for our games on Facebook and that that was a declining market. I started looking around for a new opportunity, which was when I realized what really matters in product management and in any career: relationships. A lot of my former colleagues— engineers I had worked with in several different teams in Zynga—reached out and invited me to work with them in their new ventures. I joined two of my former colleagues at Bright, heading their product team.

I continue to learn a great deal at Bright and to work with some amazing and smart people. The company operates in the big data space in a huge market (jobs), which makes it exciting and presents a big opportunity. In addition to several consumer products, I'm also overseeing an enterprise product, which is very different (especially when it comes to interfacing with the sales team and their sales cycle).

What do you think has made you successful thus far?

An open mind, a deep sense of curiosity, and constant desire to learn. You can't be afraid of going into an area that you don't know much about – you have to be comfortable getting up to speed quickly in new and potentially intimidating areas. You need to be a consummate and life-long learner. The key is to ask questions, be curious and learn from your team.

I always treat engineers, designers and other team members as equal partners, and they explain and teach me about their areas of expertise. My team showed me that coding was not scary, but interesting. It's built on logic and not nearly as complicated and mathematical as my preconceptions had made it out to be.

A sense of curiosity and passion for the product is very important as well, as is developing empathy for the customer – if the PM doesn't care about the product, it will show in the quality of product in many different ways.

As a product manager, you also have to be able to straddle multiple levels and audiences. It's like speaking different languages (human or computer ones) and being able to relate the message to people in a language that they'll understand. You need to get your team jazzed and excited about what you're building, get them to contribute and own their ideas, and also relay and manage communication up to management, sales, legal, etc. – departments that care about different aspects of the product.

You need to genuinely like being around people and working with people, bringing their best energy forward. You need to be the type of person who thrives on being in the thick of it, solving problems on the fly, and making decisions quickly without fear. You can't be the quiet person in the corner who wants to be left alone all day.

What this also means is that as a PM, you have to recognize that you will never be the best marketer, or engineer, or sales person. You have to be proficient and versed enough in a lot of these areas and have a good sense of how they fit together in the product. It's what I imagine being a conductor of an orchestra feels like.

How valuable is an MBA for product managers?

Getting an MBA just for the sake of getting an MBA is not worthwhile in tech, at least not in Silicon Valley. There are certainly certain industries where it's a prerequisite –management consulting for example—but tech is not one of them.

Having said that, an MBA from a top school is extremely valuable for the network; a lot of my classmates participate in the business and product creation process as founders of companies, venture capitalists and other key players. So I'm organically plugged into a very powerful network that will keep creating and compounding opportunities for my career whatever I decide to do next.

The other thing I loved about the MBA is the experience of spending two amazing jam-packed years learning and getting exposed to a variety of companies, industries, and people in a very intellectually stimulating and mind-opening environment. I love learning and I experienced one of my most intense personal growth phases at HBS. But I don't advertise my MBA or Harvard in my day-to-day job and interactions. Where I work and with what I do, all that matters is your ability to contribute to the success of a company. Some of my smartest colleagues haven't even gone to college, so educational credentials don't mean much.

An MBA might well be a good path for you, but you need to think carefully about what you're looking to get out of it and why you would invest the time and money in the experience.

What advice do you have for people who want to become PMs and advance in their careers?

Be curious and strive to learn something new every day. Learn from the people around you – the marketers, from engineers, from sales people, the QA guys, etc. Without curiosity and an open mind, you'll get defensive and bitter, and that's a recipe for hating your life as a PM.

Don't chase a brand. Don't just go somewhere just because it's "the" company everyone's talking about. It's funny – in the early days at Zynga, a lot of people who would apply to be product managers were the hungry, scrappy kind who had some sort of handicap. Later on, as Zynga became of the hottest pre-IPO companies in 2010 and 2011, there were a lot of people who wanted to apply to be a product manager because they had heard the company was hot and that recruiters liked product managers with Zynga experience.

Instead, think about what makes you happy. What is your style – analytical, technical, design-focused, creative? How does it fit with the culture of the organization? Does the product of the company resonate with you? Are you excited to serve the types of customers the company attracts? Do you see yourself working long days with the people you meet during your interviews? Are the problems they are solving exciting to you?

Be open minded. Product management is not an obvious and strictly defined role. A lot of companies go by for a long time without product managers and others don't have that role at all. You don't have a defined set of buttons to push and levers to pull. You have to be excited, driven, self-motivated, and compassionate. You have to be able to pitch in and help where needed – that's how you carve out your role as PM in a startup.

Finally, be able to recognize an opportunity when it shows up. It may be different than what you expect, but if it gets you in the door and one step closer to creating a product, seize it. Be scrappy and trust that it will all eventually work out.

Behind the Interview Scenes
Chapter 6

Google

A quick search online will reveal a multitude of rumors about Google's interview process and culture. Some will recount "horror" stories about Google interviewers trying to intimidate the candidate, and some will discuss the ridiculous questions that someone's friend's cousin's college roommate allegedly was asked.

The reality is so much tamer than that.

Google's interview process is much like every other company's. It begins with two phone interviews that may cover technical, cultural fit, strategic, analytical, and product design questions (though perhaps not all of these).

If you pass those interviews, you will be brought onsite for a full day of interviews. You may have an interview over lunch. If so, this interviewer generally doesn't submit feedback for you. She's just there to sit with you and field questions.

Your other interviewers will be assigned specific roles. Some will evaluate your technical skills, others your product skills, and still others your analytical skills.

Each interviewer will be evaluating you separately. Google is pretty strict about interviewers not sharing feedback on you with other interviewers until everyone has submitted their feedback. This means you don't need to worry about poor performance in your early interviews biasing your later interviewers. It also means getting "easy" or "hard" questions later on in the interview means nothing about your performance.

After the interview, associate product managers (APMs) will be given an essay to write. The essay often focuses on an area of business strategy. Be concise and to the point; beautiful, descriptive prose is not good business writing. And, of course, be sure to check for any spelling or grammatical errors.

PM candidates will be interviewing for a particular team, but many of the interviewers will not be from that team. APM candidates do not interview for a specific team since it's a rotational program.

How Decisions Get Made

Interviewers do not decide who gets an offer. Rather, each interviewer writes up feedback, which goes to a hiring committee that makes the decision.

You will be evaluated on a scale of 1.0 to 4.0 by each interviewer. This number alone doesn't mean much though, as the hiring committee will take into account whether your interviewer was a "harsh grader."

Generally, you will need at least a 3.0 average interview score, plus one strong supporter. Thus, even if all of your interviewers recommend hiring you, you still might get rejected if none thought you were outstanding.

The hiring committee is a mix of peers (other PMs), managers, and recruiters. Your interviewers will typically not be on the hiring committee. If this does happen, it's purely by coincidence.

The hiring committee makes a hire / no-hire recommendation (which is rarely overturned in later stages). If you get a hire recommendation, your packet is passed to a compensation committee and then to an executive committee to finalize your offer.

The number of stages in the Google hiring process is one reason decisions can take several weeks.

Special Focuses

Google is not as into behavioral questions as other companies are. In fact, many Google interviewers won't have thoroughly read your resume beforehand; they prefer to test your skills more directly in the interview.

Try to mix a few key points into each interview (where appropriate) since interviewers might not probe as deeply into your background. That awesome side project you did, when combined with strong performance on an analytical question, could be what you need to get a "strong hire" recommendation.

Additionally, remember the hiring committee does not speak with your interviewers directly. They must make their decision entirely from the interviewers' writeups. If there's some aspect of your background that you want to be sure the hiring committee knows, mix that into multiple interviews. This will greatly improve the odds that the committee knows about it.

Google loves questions about its own products: Which ones do you love? Which ones would you do differently? Be prepared to talk in detail about some Google products.

Google also asks a lot of estimation questions and technical questions. Be sure to brush up on both your quantitative skills and your technical skills. Don't be surprised if you're asked to write a bit of code on the whiteboard.

Estimation questions will often involve some aspects of Google products, and one of the big areas there is advertising. Pay special attention to the questions related to advertising in the case study chapter.

Associate Product Managers

Candidates with minimal experience (less than two years) generally interview for an associate product manager role. This is a rotational program, so team fit is obviously less important. The interview process is basically the same as for the product manager role.

Microsoft

Microsoft's process is perhaps the simplest and most straightforward of any company. In fact, the interview processes of many other companies appear to be a derivative of the Microsoft process: swap out the hiring committee for a hiring managers, and voila.

At Microsoft, candidates usually start with one or two phone screens. If you're a college candidate, this interview might take place on your college campus.

Sometimes a recruiter does an initial screen. That's no reason to drop your guard; just because it's by a recruiter doesn't mean it can't be challenging.

Candidates with work experience often have a technical screen as well.

After the phone screens, candidates are flown out to Redmond for a full day of interviews. Candidates will generally interview with a specific team. The requirements might change from team to team, so a candidate who is a great fit for one team might be a lousy fit for another.

If you interview with the hiring manager or a more senior person at the end of the day, this is a good sign. It often means you have gotten a "hire" from a technical point of view, and they often are just testing cultural fit or fit with Microsoft as a whole. It may also be that they are teasing out some final points people were on the fence about. This interview is called the "as app" ("as appropriate") interview.

How Decisions Get Made

After your interview, the interviewers for a team discuss your performance via email or an in-person meeting. They make a decision and pass it back to a recruiter who will get together an offer packet, if necessary.

In some cases, candidates discover they will receive an offer before they have even left Microsoft's campus. This might seem shocking, but remember: all that interviewers need to do is provide their feedback and potentially discuss it quickly. If all your interviewers happen to be available that day to chat, decisions can happen very quickly.

If this doesn't happen to you though, don't be disheartened. Being able to make rapid decisions has to do with your interviewers' schedules and the number of other candidates in the pipeline, not your performance.

Special Focuses

Microsoft in particular likes behavioral questions and product design questions.

In product design questions, pay attention to the details and be sure to ask probing questions. Microsoft interviewers often enjoy testing how you handle ambiguity. They might, for example, ask you to design a pen and not mention that it's a pen for astronauts. They want to see that you ask a lot of questions to understand the customer before running down some path.

Extra Goodies

Now that we've generalized about what Microsoft does, we should also leave you with this advice: be careful about generalizations at Microsoft, particularly with respect to hiring. What your buddy experienced at Microsoft might have little to do with what you'll experience.

Microsoft teams hire mostly independently. One team might want deep technical skills and therefore demand that you write some pseudocode, while other teams might want to test your design skills. It's all over the map.

Facebook

Facebook was built on a hacker culture, and this shows in their culture. Facebook wants PMs who show this entrepreneurial drive.

Like other companies, PM candidates start off with one or two phone interviews. The phone interviews often cover behavioral questions, and may also ask why you are passionate about Facebook. The best answers may include anecdotes about real-world experiences with Facebook or a desire to have a big impact and work with smart people.

If you do well on the phone interviews, you'll be brought in for four to five onsite interviews. Each interviewer wears a specific "hat":

- **Technical and Logical:** You will be asked some quantitative questions, especially around metrics and implementing experiments. If you have a technical background, you might also be asked to code (although Facebook has relaxed this requirement more recently).

- **Design:** These will include typical product design questions. Additionally, if you've built anything, they may ask to see it. Think about what services and apps you like and why you like them.

- **Futurist:** These include questions like, "What is the future of TV?" Show that you can reason about the future. You don't just want to talk about symptoms of the future; you want to think about what will fundamentally change to get that way and what effect those changes will have. You want to be a storyteller.

- **Guru:** For experienced roles, they'll ask about your core strengths. They want to check your sense of self to see if you understand what you're good at.

Entry-level PMs generally get hired at Facebook as generalists rather than for a specific team.

Experienced PMs mostly interview for a specific team, but Facebook still expects these candidates to be generalists at heart. They will interview with some team members and some non-team members.

How Decisions Get Made

Facebook interviewers do not directly make offer decisions. Instead, interviewers submit written feedback to a hiring committee which is a mix of peers, managers, and recruiters.

You will be evaluated in several categories during your interview. The hiring committee will need to see strong performance on all of these categories to extend an offer.

Special Focuses

Typically, Facebook will ask PM candidates to code. They understand you might not have coded in a long time, and that will be taken into account. What they're looking for here is someone who can think like a software engineer. Do you understand, more or less, how to break down a problem into steps? Knowing the ins and outs of data structures and algorithms is generally not expected, but you should know some basic ones like hash tables.

Apple

People joke that Apple is a cult, and perhaps there's some truth to that. Apple does really value culture fit. This is reflected in their interview.

Apple hires for specific teams, not for the company as a whole. Apple's interview process kicks off with two phone interviews. After that point, you are brought onsite for in-person interviews.

Some teams stick with the standard four to five hour (or so) long interviews with members of the team. Other teams, however, may give you as many as twelve 30-minute interviews. These teams apparently value culture fit so much that they want you to meet with a lot of people.

Your interviewers will come from a variety of roles: other PMs, designers, engineering managers, an executive (e.g., Junior Vice President of Product Managers). In many cases, a hiring manager will interview you over lunch, but they might also ask someone to fill in for them if they're busy.

Part way through the day, the interviewers might check in with each other to ensure they're following the right process.

How Decisions Get Made

After your interview, the hiring manager and team will meet together to make a decision.

Special Focuses

Apple believes passionate employees make good employees, so they want people who are passionate about the company and its products. Expect a lot of questions about why you want to work at Apple. Have a good pitch ready.

Similarly, you should know Apple's products well. Be prepared to describe what you love about them and what you think could be improved.

Amazon

Amazon candidates start off with two short phone screens. These are just 30 minutes, and they typically don't drill too deeply into your skills. They're mainly looking to understand your background to see if you should be brought onsite.

The onsite interviews consist of four to six in-person interviews, each about an hour long.

Interviewers will be looking to see how you match up against Amazon's 14 leadership principles (see: Amazon Leadership Principles, pg 358), with each interviewer covering two to three principles. If an interviewer doesn't feel that she did an adequate job covering one, she might ask another interview to follow up.

Of these 14 principles, an ability to get things done ("bias for action" and "deliver results") and customer obsession are particularly important.

One of your interviewers will be the "bar raiser." The bar raiser is a special interviewer from another team. This interviewer is tasked with "raising the bar" and ensuring you are better than 50 percent of Amazon PMs. He is often easy to pick out from your interviewers: he's the one brought in from another team.

The bar raiser is also often the person who will challenge you the most. For example, he might be testing the "having backbone" value. Do you have a position that you can back up while respectfully disagreeing? Don't be surprised if the interviewer continues digging into something until you've given a satisfactory answer.

You will also likely interview with a hiring manager.

How Decisions Get Made

After your interview, your interviewers will meet to discuss your performance. The bar raiser is responsible for the interview process and gets veto power. The hiring manager also gets veto power; it's her team, after all. This means you need to impress both the hiring manager and the bar raiser (and ideally everyone else, too).

Special Focuses

Amazon tends not to focus too much on technical skills, although some technical aptitude might be expected in more technical teams like Amazon Web Services.

What the company cares about more are your business skills and background.

All of Amazon's leadership principles are important, but the Customer Obsession one is especially important. When in doubt, do what's right for the customer (even if it isn't the right "business" decision).

Many Amazon questions deal with pricing specifically, so make sure you think about how different Amazon products (e.g., Amazon Prime) are priced. Think about what you would change.

Amazon interviewers like to dig deep into your resume. That line you have about how your feature boosted efficiency by 30 percent? You'd better back that up. Be prepared to justify exactly how it did this and exactly how you measure this impact. Hand-waviness won't cut it.

Finally, these leadership principles are not a joke. If you pay attention and know the leadership principles well, you might recognize which one an interviewer has in mind with a particular question. You can then address it directly. Better yet, prepare for this; review your resume with these leadership principles in mind.

Yahoo

Yahoo recruits for both the product manager and the associate product manager roles. The procedures for these interviews are mostly similar, but they have a few differences.

Both product manager and associate product manager candidates start off with one or two phone screens. Successful candidates are then brought on for a full day of interviews.

In the onsite interviews, PM candidates interview with people from multiple roles and levels. They should have at least three interviews with fellow PMs of the same level or higher, plus at least one person from a different team. One of your interviewers will also be the hiring manager.

In APM interviews, candidates do not interview with the specific team. After an offer is made and accepted, you will be asked your team preferences. You will then be notified of your team one to two weeks before you start.

How Decisions Get Made

After your interviews, each interviewer submits written feedback to the hiring manager, who then puts together a packet. If the team is feeling positive about the candidate, the packet is sent to the hiring committee and then to executives for final approval.

For APM interviews, offer decisions are made by the APM steering committee, then reviewed and finalized by the executives.

In either case, Yahoo looks for people who are passionate, high energy, and capable of getting stuff done and launching products.

Special Focuses

Yahoo is looking for deeply technical PMs, so you should expect to prove that you have a solid technical foundation. You need to show you can communicate with engineers, but it's unlikely you'd be asked to code.

You should also expect product and analytical questions. Try to have a framework and a specific point of view.

Twitter

Twitter's process starts with a hiring manager who does a general phone screen and then matches you with a specific team. You will do one or two phone screens before being brought onsite.

In the onsite interview, you may go through as many as seven interviews that are 45 minutes each. Your interviewers will be a mix of peers (fellow PMs) as well as people whom you might work with, such as engineering managers, tech leads, or people from the support team.

How Decisions Get Made

Each hiring manager does things a bit differently. However, Twitter generally only extends an offer when someone is a "slam dunk." They want someone who brings something new to the team.

Special Focuses

Twitter really wants people who have done their homework and love Twitter. Applying to Twitter just for its brand name (as oh-so-many people do) won't cut it. Instead, you should play with their technology and really understand it. You need to "get" Twitter, not just be a casual user. What do you think is really cool? How does it work? What would you do if there were an issue? You should be obsessed with creating a great experience for the user.

Twitter also wants people who can handle change, since Twitter is growing rapidly. You should be willing to wear many hats, be good in stressful situations, and have excellent interpersonal skills. Behavioral questions are very important.

Twitter PMs will generally not be asked coding questions, but they may be asked how to technically design a product. You should understand concepts like preloading and calculating on the fly.

Dropbox

Drew (Co-Founder and CEO) tells each new hire that their #1 job is to recruit other talented people. Given this, it's not surprising that almost half of new hires are referrals and the recruiting team is very active in sourcing candidates from existing PMs.

While the interview process is still evolving, the PM candidate profile is very well defined. Dropbox is looking for people with a technical background that also have experience as a startup founder or who have demonstrated substantial accomplishments as a PM at an established company.

Once you're in the door, you go through two phone screens with other PMs who ask typical PM questions about your favorite products and potential improvements.

If you are invited onsite, you will typically face four interviews with PMs, engineers, and product designers. These will include PMs who ask product questions, engineers who go through a technical screen, and product designers who may ask you to design a new product workflow on the whiteboard. Either that day or soon after, you'll also interview with Arash (Co-Founder and CTO) who asks product questions related to Dropbox and also screens for cultural fit.

Depending on current needs and candidate backgrounds, PMs may be hired for a specific team or simply as a generalist to jump into a specific team soon after starting.

How Decisions Get Made

After the interviews, the interviewers will debrief together to make a decision. The recruiter is in touch with the interviewers throughout the day, so if you don't pass the initial interviews, you may not speak with Arash. The bar is high and Dropbox will take their time to find the right candidate.

Special Focuses

One of Dropbox's engineering values is to "sweat the details," and it applies to the PMs as well. Be prepared to think through all the edge cases of your product answers and designs in detail. Cultural fit is also extremely important, but they don't ask any special questions to screen for it.

Dropbox has a very focused set of products, so be familiar with all of them and think through what you would do if you were a PM there.

Resumes
Chapter 7

It's not your experience that lands you an interview; it's how your resume presents that experience. Even the best candidate in the world won't get an interview with a lousy resume. After all, a company wouldn't know that she's the best candidate in the world.

And, in fact, many great candidates do have lousy resumes. They lack perspective about their own experience, get confused by out-of-date advice from career counselors, don't back up what they've done with specifics, or write generic lofty statements that end up meaning nothing to the resume screeners.

A bad resume is an issue for any job, but especially so for product management roles. Communication is an important PM skill, and your resume is one clear way to demonstrate that. A PM who can't express her skills and accomplishments in a clear, concise, and effective way is a bit worrisome. Much more so than in other roles, you'll be judged for the quality of your resume.

The 15 Second Rule

You know what a resume is, but do you truly understand how it's used? (Wait! Don't skip over this section! You actually need to understand this to write a good resume.)

A resume isn't read; it's skimmed. A resume screener will glance at your resume for about 15 seconds (or maybe less) to make a decision about whether or not to interview you.

This forms the guiding principle of resumes. A resume should be optimized for that 15-second skim.

Let that soak in. We will come back to this principle again and again.

The Rules

Every rule has its exception, but they're called rules for a reason. Proceed carefully if you think one of these rules doesn't apply to you.

Rule #1: Shorter is Better

Imagine I wanted to tell you as much about myself as I can, but your attention span is only 15 seconds. Should I give you my 300-page autobiography? Or my condensed one-paragraph version?

The 300-page version will have a lot more information, but that doesn't matter. In 15 seconds, you'll only have time to read the first paragraph. I'd be lucky if you learned where I was born in that time. Although I offered more information, you actually learned a lot less about me.

A long resume is like that. It takes all your best content and then mixes in less important information, leaving the resume reader with a worse overall impression of you.

It's best to stick to just the highlights.

Implementing This Rule

A good rule of thumb is to limit your resume to one page if you have less than 10 years of experience. At more than 10 years, you might be able to justify 1.5 or 2 pages, particularly if you've held many different jobs.

Before you say you can't possibly fit everything you've done on one page, you're right; you can't. However, you can fit the most important things on one page. You might have to be more concise, but that's a good thing. This means that you're sticking to the highlights.

When you're thinking you need more space for a particular role, ask yourself what about that role is most important. Is it the fact you were a coder at one point? Is it the impact you had in reducing the company's costs? Is it just the name of the company?

Focus on what is important, and leave out the rest.

Also, if just a few lines force your resume onto another page, find a way to trim down your resume. A resume that just barely goes onto a second page suggests a poor ability to prioritize.

Rule #2: Bullets, Not Blobs

Many people live by a rule of "talk more about what's more important." There is some wisdom to this guidance, but it can backfire on a resume. The longer a chunk of text is, the less likely a resume screener is to read the resume.

Blobs of text—that is, bullets or paragraphs that are three lines or longer—tend to not be read. Keep things short.

Implementing This Rule

Read through your resume. Anything that's three lines of text or more should be condensed. Additionally, you should aim to have no more than 50 percent of your bullets expand to two lines. That is, at least half of your bullets should be just one line, with the remainder being two lines.

Depending on the situation, this might require alternative word choices, or it might require cutting out some of the details. The impact of your work often matters more than the details, so it's okay to skimp here.

Additionally, if just a few words of a bullet cause it to flow to the next line, trim it. You will waste space otherwise.

Rule #3: Accomplishments, Not Responsibilities

People don't care what you were told to do; they care what you did.

Responsibilities are about what you were told to do. Statements outlining these offer only a broad, fluffy overview of what you were supposed to do in your job. They don't make it clear if you actually had an impact. Moreover, your responsibilities are often pretty obvious. We know, broadly speaking, what a product manager or a software developer would do at a company.

Instead, you want to focus on your accomplishments. Prove to the resume screener you had an impact.

Consider the difference between these two bullets:

- **Responsibility Oriented:** Design features for Amazon S3 and oversee development of the features across software engineers and testers.

- **Accomplishment Oriented:** Designed the SS Frontline feature, managed its development, and led its integration across three products, leading to an additional $10 million in revenue.

While the first bullet gives some some information about what you did, the reader won't walk away saying "You were a success because_____." Make your resume look more like the second bullet. That shows success.

Implementing This Rule

Using the present tense is a good tip-off that you've listed a responsibility. It's difficult for something you accomplished to be written in the present tense.

Making your resume accomplishment oriented goes beyond that, though. After all, if you took the earlier example bullet and converted it to past tense, it still wouldn't be a true accomplishment.

Instead, list the concrete ways you had an impact. Focus on the impact itself; the "what" more so than the "how" (although both are important).

As much as possible, quantify your accomplishments. How much money did you make for your company? How much time did you save your team? By how much did you improve customer retention? An estimation is okay here.

If you have an existing resume, it might help to start from scratch with one of these questions in mind:

- What are the five things you are most proud of?

- What would your team say are the five most impactful things you did?

The answers to these questions should form your bullets.

Your responsibilities should generally be clear from your specific accomplishments and from your job title. However, if you feel you must explain your general responsibilities, a good place is immediately under the job title and in italics, so as to separate it from your true accomplishments.

Rule #4: Use a Good Template

Every few months, some website or blog publishes a list of "amazing" resume designs which, undoubtedly, a bunch of job seekers attempt to copy. Such resumes use infographic-style charts or mockup a resume in the style of the Amazon.com homepage or the iOS homescreen.

These resumes are cute, they might show some degree of creativity, and they might even grab someone's attention. But unless you're one of the lucky few to garner some media attention for your flashy design (or you're applying for a designer position), a resume template like this will generally hurt you.

Many hiring managers hate these graphical resumes because it's hard to learn much about you. Information isn't presented in a clear way, and the information that is presented takes up way more space than necessary.

A good resume template won't make your friends "ooh" and "ahh." It probably won't be flashy or particularly creative. But it will get the job done—which means landing you an interview.

Implementing This Rule

A good resume is reasonably compact and quickly showcases your highlights.

Look for a resume template with the following:

- **Two or three columns, one for company names and the other for jobs titles.** You want to make this information very easy to pick out, especially if you have a top company on your resume. Location and dates are considerably less important. They need to be there, but they don't need to leap out at the resume screener.

- **No left column dedicated to headings.** Many resume templates use the left side of the page for headings such as "Employment" and "Education." This looks attractive, but can waste 20 percent of the available space.

- **Limited text stylings.** Too many fonts, sizes, casings, and colors can be distracting.

- **Reasonable use of whitespace.** Too much whitespace wastes space. Too little can make your resume difficult to read and can suggest that you're not good at prioritizing.

- **Reasonable font size and margins.** You want something that's easy to read while not being wasteful with space.

- **Bullets.** Blocks of text look pretty (particularly on a graphical resume), but will be skipped over.

Most resume templates meet these criteria. You can find some samples at www. crackingthepminterview.com.

Rule #5: Don't Skip the Best Stuff

In theory, this is obvious. Of course you shouldn't leave the best stuff off your resume!

In practice though, many candidates ignore this final rule. They leave something out because they didn't feel it was "appropriate" for a resume for <insert strange reason>. This is so common and so important that we've listed it as a rule.

For example, Jessica, a product manager at Amazon, was applying for other PM jobs at Amazon and other companies. After multiple rounds of resume editing and feedback, her resume was almost perfect—except for one detail. She'd neglected to mention that she had, on the side, launched a gaming company, hired multiple developers and designers, and overseen the development of a game. The combination of this entrepreneurial effort and the Amazon job are basically her golden ticket into any PM interview.

Why didn't she include it? Because, due to some medical problems, she hadn't yet launched the game. She figured that you couldn't list it until you were "done."

In similar situations, other people have given a variety of reasons: "It was for a class." "It wasn't an official class project." "We haven't finished yet." "We didn't get many downloads."

None of these reasons are sufficient to exclude something from your resume. If it helps you, list it.

Implementing This Rule

Ask yourself: what did you *not* include? Are there projects you've done (on your own, for school, for a friend's company, for a hackathon, etc.) that you haven't listed? Any relevant hobbies? Or interests which have some interesting accomplishment (e.g., completing a triathlon)?

There are no hard and fast rules about what belongs on your resume and what doesn't. If it makes you a more interesting or more attractive candidate, include it.

Attributes of a Good PM Resume

Employers want PMs who have technical skills, love technology, possess initiative, are leaders, and will have an impact. A resume is a chance to showcase these parts of your background.

It's more than that though. A resume is itself a product. It makes a statement about your communication skills, design skills, and your ability to put yourself in the "user's shoes."

Think about your audience: What do they care about and how will you demonstrate that you have those things? For example, if you work for a company that's not well known, can you concisely describe what the company is on the resume? Is there a way you can establish credibility, such as mentioning who funded the company? This is your chance to show off your "product design" skills.

For many PM positions, it will be important to demonstrate these skills:

- **Passion for Technology:** If you have technical skills or have worked at a tech company, this will probably show enough passion for technology. If you don't have these things, find some other way to get involved with technology. You could start learning to code through online courses, build your own website, or even outsource development of a project.

- **Initiative:** You could show initiative through a club at your university, a new employee training program at your startup, or even a monthly dinner for people interested in technology. Show these experiences on your resume.

- **Leadership:** If you've managed people in some capacity, show this. This experience could include mentoring / managing an intern or being the president of a club or organization.

- **Impact:** Show that you've had a positive impact in your prior roles. Be clear about what you've personally driven, since your team's accomplishments are much less relevant than your own. Explicitly state what you've built, created, led, or implemented. Avoid weak phrases such as "worked with" and "helped with."

- **Technical Skills:** If you have programming skills, list these programming languages in a "technical skills" section. This will suggest some degree of proficiency. Ideally you will also have specific projects to list.

- **Attention to Detail:** This is more about what you don't do than what you do. No spelling or grammar mistakes. Consistency in ending bullets with periods (periods are optional, but you need to be consistent). Correct contact information.

Go through your resume and look for signs of each of these attributes. If you're missing some of these attributes or skills, find ways to acquire them and then add them to your resume.

What to Include

Your resume should obviously include your work experience and education. What about all the other little details?

Objectives: No

Objectives are almost always a waste of space. Let's dissect this example of a typical objective:

> "Experienced technical leader with a bias for action seeking product management role in a fast-paced, growing company."

The "bias for action" part is subjective and a claim anyone can make. The "experienced" description would be clear from the candidate's resume. The description of the company doesn't help either. You're applying to this company; you are, by definition, interested in it, whether it's fast paced or not.

Objectives are just a verbose way of describing the role you're applying for. There is no need to state what will already be clear from your application.

Summary: Rarely

With few exceptions, a summary is rarely useful. If your resume is sufficiently concise, it already *is* a summary. There's no need to re-summarize it in paragraph form.

Moreover, most summaries are laden with fluffy, subjective personal descriptions such as "dynamic" and "action-oriented." These carry little weight in the eyes of the reader.

Occasionally, summaries can highlight specific accomplishments or responsibilities that might otherwise not jump out at the reader. However, this case is unusual. A proper design can almost always make your highlights readily apparent.

Skills: As Needed

You might want to include a skills section on your resume, particularly if you have programming skills or experience with design software. Skip obvious skills, such as Microsoft Word. Listing this as a skill communicates only that you know how to open a file, edit it, and save it. Everyone knows how to do that.

Awards: Yes—And Make Them Meaningful

You should list awards you've received. Even ones that don't seem directly applicable to the skillset are often relevant in showing success, hard work, or creativity.

Many candidates list their awards but fail to make their awards meaningful. They list an award like this:

*   John R. Robertson Memorial Prize (2013).

A resume screener has no idea what to make of this. What is the John R. Robertson award for? How select is it? What did you do to win it?

Ideally, your resume should describe what the award is for *and* how selective it is. The award above, for example, might be listed as:

*   John R. Robertson Memorial Prize (2013): Placed 1st out of 75 students in business plan competition. Entered business plan for low-cost solar heated pools.

This establishes both relevance and selectivity.

Activities: Sometimes

Depending on what the activity is and what you've done with it, activities and interests may be useful.

The more relevant the activity is, the better it is to list it. Technical activities and creative work can be easily seen as applicable.

Even if the activity itself isn't particularly relevant, what you've done with it might be. For example, running is probably not applicable to your job application (unless you're applying to a fitness-related company). However, if you've completed a marathon in 17 states with a lifetime goal of covering all 50, then it might be relevant in showing determination—and in just making you "interesting."

Think about activities the way you would other work on your resume. Try to back each up with a concrete accomplishment. If you're just another person listing "mountaineering" on a resume, it usually won't mean too much.

Projects: Yes

Projects are incredibly important. In fact, projects are probably the second most important thing, after work experience.

If you have any side projects, list them on your resume. Describe what the project is that you built and what metrics of success you have.

For example:

- **Snakes and Ladders** (iOS game): Designed UX for multi-person turn-based iOS game, and hired outsourced development team to implement game. Achieved 10,000 free downloads in first month with 10% conversion to paid version.

This shows the leadership responsibilities you took on and quantifies your impact.

Website URL: Yes

If you have a website or blog, you should include the URL on your resume.

If you don't have a website, consider building one. Your website should provide your resume, as well as additional details about your projects (such as screenshots). It can also list articles you've written, press you've gotten, lectures you've given, and other related interests.

Some basic personal information is fine, but keep the website mostly professional. Whether you list it or not, employers will likely look at it.

Social Media Accounts: Maybe

If you are active on social media about technology or work-related issues, your social media accounts could prove valuable. Make sure to scrub any old posts that might reflect poorly on you.

College / University Details: Sometimes

The further you are from college graduation, the fewer details should remain on your resume. Exactly when to remove items depends on what the activity was. Use the following as general guidance:

- **Club Membership and Other "Participatory" Items:** Simply being part of a club doesn't say anything about you. These can be removed upon graduation, if not earlier.

- **Programming Projects:** If you have programming projects, these can stay on your resume for about 2 - 3 years after graduation. If you can replace them earlier with more interesting projects, that's even better.

- **Substantial Leadership Positions:** If you were the president of a club or had major accomplishments as a leader, you could justify keeping these on your resume for 2 - 5 years after graduation. Simply being "VP of Marketing" for a club doesn't say much and can be removed fairly quickly.

- **Founding Accomplishments:** Founding a club, charity, sport, or another major activity shows you have initiative and that you get things done. Therefore, you can justify keeping it on your resume for a bit longer—possibly 5 - 10 years, depending on how significant the accomplishment was.

- **Awards:** This varies wildly depending on the awards. An extremely impressive award could conceivably stay on your resume for 10 years (or even longer). Less impressive awards, such as winning third place in a college programming competition, should probably be removed within about two years.

This guidance is merely a rule of thumb, and it might not be the right advice for you and your resume. How long is appropriate for you depends on what the item shows, how important that skillset is, how else you show that, and what the opportunity cost of including that item is. For example, if your resume paints you to be a stereotypical geek and you're applying for a product manager job, being a member of a standup comedy group might stay on your resume for five years or more.

Generally, your education will go at the top of your resume if you're in school, but otherwise your work experience comes first. Rather than thinking of this as a hard and fast rule, ask whether you would prefer companies see your company or your school first. If your work experience is much better than your education (or vice versa), it might make sense to break this "rule."

GPA

For the first few years out of school, you should list your GPA if it's above a 3.0 / 4.0. If you are more than about five years out of school, the minimum goes up to around a 3.5. The reason for this is simple: don't list something from a long time ago unless it really makes you stand out.

If you have an exceptional GPA, there's no last possible timepoint for listing it. You can always tack it onto the bullet listing your major. Most people probably won't care, but it doesn't take any extra space. The opportunity cost is zero.

If your university computes GPA on something other than a 4.0 system, it may be hard for other people to understand what your GPA means. In this case, try to translate your GPA into something more meaningful. For example, you could translate your GPA by appending a line like ("Equivalent to a 3.3 / 4.0") or you could list a class rank or percentile.

Online Courses and "Extracurricular" Education

If you've been taking online courses, you can list these on your resume. Doing so shows a passion for learning as well as some expertise in the subject, both of which are good things.

The tricky thing is how to list an online course. If the course is significant enough, you can list it under "Education," but more likely you'll put this in an "Additional Information" section.

Find ways to make these classes sound more legitimate. If you have a (good) grade, you can list that. Or, if you have completed interesting projects for these courses, that will help as well.

Real Resumes: Before & After
Chapter 8

Even when a resume was good enough to land someone a PM role, it can usually be improved. To give you a sense of what kind of improvements you can make, we're sharing examples of the real resumes that landed these jobs, along with examples of how they could be made even better.

Note: The "before" resumes shown here match what the candidate submitted to get a PM role, although they have been anonymized in some cases. Anonymizing resumes involved changing names and key details, but keeping the facts roughly equivalent. For example, rather than saying that a candidate went to Columbia University, we might instead say Cornell University.

Richard Wang (Anonymized)

Richard Wang received offers from Dropbox, Google, and Uber in 2013 using the resume below.

The Original Resume

EXPERIENCE

ImaginiNow
Head of Product (January 2013 - April 2013)
- Led product team to help single people meet friends of friends
- Achieved 90% increase in requests to meet people, 97% increase in signups and 23% increase in viral invites
- Drove product strategy based on company vision, user feedback and metrics; cut features in half to increase focus
- Developed feature ideas into detailed specs and collaborated with designer to create mockups
- Defined, implemented and analyzed metrics using Mixpanel to measure product success
- Prioritized work for weekly sprints and coordinated with engineering and design teams to track progress
- Established new channels for customer development including usability focus groups and a user advisory board
- Applied lean startup methodologies to design experiments, build minimum viable products and iterate rapidly

Connecto
Founder (2010 - 2013)
- Created a social networking app to connect professionals for job opportunities and sales leads
- Raised $500K from investors including Jason McMillan (initial investor in Vault.com) and Infinity Venture Partners
- Drove product vision, design and execution based on customer needs and competitive landscape
- Wrote feature specs, designed user experiences and created mockups using Fireworks, HTML/CSS, and Axure
- A/B tested landing page design, email copy, Tweets and Facebook posts to increase conversion
- Defined and analyzed key performance metrics to optimize user engagement
- Developed site using Ruby on Rails, HAML, jQuery, CoffeeScript, SCSS, Twitter Bootstrap, PostgreSQL and MongoDB

Microsoft

Project Manager (2005 - 2009)

- Senior member of the project management team that guided development of Windows
- Led 20% of cross-functional projects to develop new technologies for Windows: wrote engineering plans, recruited ten-person development teams, tracked progress and presented results to Senior Vice President of Engineering
- Coordinated with 30+ engineering teams to unblock progress, prioritize tasks and meet milestones
- Communicated project updates and goals across 860+ person team; produced status reports for SVP of Engineering
- Worked with marketing to develop product demo for Microsoft Build Developers Conference; met weekly with Senior Vice President of Marketing for feedback
- Organized internal tech talks program to evangelize new technologies; daily attendance of 300+ people
- Developed project management tools using C++, C#, and SQL that were used across the organization
- Debugged technical OS issues alongside engineers to unblock daily builds

Madison Assisted Living Facility

Project Lead (2004 - 2005)

- Led 15-person team to design and implement web-based, touch-screen medical log

EDUCATION

Kellogg School of Management, Northwestern University

Master of Business Administration; Innovation & Entrepreneurship Major (2009 - 2011)

University of California, Santa Barbara, School of Engineering

B.S, Computer Science & Engineering Major (2001 - 2005)

ADDITIONAL INFORMATION

- Interests: Adventure travel (hiked Inca Trail, sailed Gulf of Mexico, ice climbed in Adirondacks), photography and Krav Maga

Assessment

Richard's resume is actually fairly good, compared to many other resumes. However, it could still be stronger.

Most importantly, Richard could make his resume more accomplishment-oriented. He should take each job and ask himself, "What are my five biggest accomplishments?" These should form the bullets of his resume.

For example, look at the bullets under his most recent role. They are vague statements such as "drove product strategy" and "prioritized work." These aren't truly accomplishments; they're just responsibilities worded in the past tense. Until you've shown an impact, it's not really that meaningful.

Additionally, Richard has many bullets under each job, making each bullet less impactful than if the list were restricted to just the highlights.

If he's doing this to take up space on his resume, he'll need to find some other side of his background to offer. Perhaps he took on some interesting roles during his MBA or has some projects he's done on the side.

Finally, he should consider listing his programming languages under "Additional Information." He's got the space (once he trims down his bullets), and this list would be a good way of quickly showing an interviewer that he is very technical.

With that said, here is what an improved resume might look like. We've offered explanations of what some of his companies do and offered more concrete accomplishments with his bullets.

New and Improved

EXPERIENCE

ImaginiNow
Head of Product (January 2013 - April 2013)
ImaginiNow is a Sequoia-backed social networking website with 5+ million members which helps single people meet friends of friends.

- Led product team of ten engineers and revamped engineering processes, implementing weekly code sprints, agile methodologies, and code reviews.
- Achieved 90% increase in requests to meet people, 97% increase in signups, and 23% increase in viral invites.
- Led project to redesign website and developed new data-driven approach, which resulted in a 97% increase in signups and a 90% increase in requests to meet people.
- Increased viral invites by 27% by designing a new feature to connect with existing Facebook and Twitter friends.
- Pared down feature list by 50% to enable more rapid development of critical products.
- Designed and led three usability studies with more than 100 people participating.

Connecto

Founder (2010 - 2013)

Connecto is a social networking app to connect professionals with job opportunities and sales leads. It peaked with 100,000+ members and ten million pageviews per month.

- Successfully led fundraising efforts for company, raising $500K from investors including Jason McMillan (initial investor in Vault.com) and Infinity Venture Partners.
- Built initial prototype / minimum viable product which allowed users to quickly search through their LinkedIn connections based on a "need."
- Led team of four (two engineers, one designer, one tester) through two moderate pivots, redefining strategy and shifting team to new vision.
- Developed company dashboard to enable tracking of user engagement.
- Developed website along with two engineers using Ruby on Rails, HAML, jQuery, CoffeeScript, SCSS, Twitter Bootstrap, PostgreSQL, and MongoDB.

Microsoft

Project Manager (2005 - 2009)

- Senior member of the project management team that guided development of Windows.
- Led cross-functional teams of 10 - 20 employees to develop new technologies for Windows: debugged issues to unblock daily builds, managed project schedules, and coordinated project goals.
- Launched internal tech-talks program to evangelize new technologies and build daily attendance to an average of 300+ people.
- Managed weekly status reports for SVP of Engineering by synthesizing information from four teams of 30 engineers.
- Built new project management tool using C++, C#, and SQL and rolled out usage across team of 25 project managers.
- Developed product demo for Microsoft Build Developers Conference (5000+ attendees).

Madison Assisted Living Facility

Project Lead (2004 - 2005)

- Led 15-person team to design and implement web-based, touch-screen medical log.

EDUCATION

Kellogg School of Management, Northwestern University

Master of Business Administration; Innovation & Entrepreneurship Major (2009 - 2011)

- **Beer Aficionados** (President / Founder): Founded beer club and grew club to 100+ members. Raised $20,000 in sponsorship from local breweries.

- **Launch Accelerator:** Accepted into Kellogg's Launch Accelerator, an incubator for startups with 15% acceptance rate.

University of California, Santa Barbara, School of Engineering
B.S, Computer Science & Engineering Major (2001 - 2005)

ADDITIONAL INFORMATION
- Programming Languages: Ruby on Rails, C++, C#, SQL.
- Interests: Adventure travel (hiked Inca Trail, sailed Gulf of Mexico, ice climbed in Adirondacks), photography and Krav Maga.

Paul Unterberg

Paul worked his way up from tech support into a product management role at a top startup. Here is his most recent resume.

Original Resume

Software Product Manager

Areas of Expertise: Minimum viable product, market analysis (product/market fit), Agile development, wireframing/prototypes, 10x productivity in startup environments.

PROFESSIONAL SUMMARY

Pricelock, Inc., Redwood City, CA (2010–Present)

Senior Product Manager

Goldman Sachs and Artiman Ventures funded startup. Ownership of multiple web-based financial service, risk management and energy trading products. Responsible for design and specification of new products and enhancing existing products.

- Launched online marketplace for energy traders, transacting $2 billion+ worth of energy products
- Increased the productivity of development team by leading in Product Owner role
- Conceived services and products that helped company close fundraising from new investors

TechExcel, Inc., San Francisco, CA (2003–2010)

Associate Director, Product Management (2006–2010)

Product Manager (2004-2006)

Managed product organization. Created product strategies for TechExcel's flagship product line, worked with customers to define requirements and features, directed outsourced and globally distributed development teams. Evangelized technology solutions to prospects, customers, analysts, and internal teams.

- Introduced Scrum, significantly improving product development
- Created systems to empirically prioritize and balance the customer's and TechExcel's business needs
- Brought new products to market and doubled revenues from software renewals

Senior Solutions Engineer (2003–2004)
Provided technical consulting, training, and ongoing support to industry-leading customers
- Consulting for deployment of ALM and CRM practices and software; achieved the highest level of customer satisfaction on every deployment (5 out of 5 customer rating).

Microsoft Consulting/San Francisco Unified School District, San Francisco, CA (1999–2003)
Senior Database Architect
- Administered multi-platform network environment overseeing a team of DBAs that supported 100+ database systems for all SF schools.

EDUCATION

BS - Computer Science, San Francisco State University

LANGUAGES AND TOOLS

PHP, SQL, Visual Basic, C, JavaScript, HTML, CSS, Excel, Balsamiq, PowerPoint, Apache, MySQL, Linux, AWS, Oracle, Marketo, Google Analytics, Wordpress, Photoshop, Fireworks, SalesForce

Assessment

The major gap in Paul's resume is specifics. He'll assert something such as "brought new products to market," but it's not really clear which projects those are. Similarly, he'll say that he "increased the productivity of development team," but he doesn't explain exactly how he did that.

There's another issue that's more minor but still worth correcting: formatting. He's bolded the job titles as though to say, "Hey! Look at me! I'm a product manager!" This isn't a particularly interesting detail; many job applicants will be product managers. Instead, he should highlight his companies.

The "areas of expertise" section won't hurt Paul, but it also won't help much either. It's better to demonstrate these skills through accomplishments.

Some projects or activities outside of work would also be an improvement. If he has these, he should list them.

Here's how an improved resume could look.

New and Improved

PROFESSIONAL SUMMARY

Pricelock, Inc. (Redwood City, CA. 2010 - Present)
Senior Product Manager
Goldman Sachs and Artiman Ventures funded startup. Ownership of multiple web-based financial service, risk management and energy trading products. Responsible for design and specification of new products and enhancing existing products.

- Launched online marketplace for energy traders, transacting $2 billion+ worth of energy products.
- Conceived of and owned a proof-of-concept predictive trading feature, which demonstrated ability to increase profit by 25%. This was a key driver in successfully closing a $10 million fundraising round.
- Established and rolled out better development practices for team in order to improve productivity and product quality, including readability guidelines, timeline estimates, Agile development, "no-meeting Wednesdays," and regular bug triaging.

TechExcel, Inc. (San Francisco, CA. 2003–2010)
Associate Director, Product Management (2006–2010)
Product Manager (2004-2006)
Managed product organization. Created product strategies for TechExcel's flagship product line, worked with customers to define requirements and features, directed outsourced and globally distributed development teams. Evangelized technology solutions to prospects, customers, analysts, and internal teams.

- Owned and launched data analysis tool to enable doctors to more accurately track clinical trial data, leading to a 95% reduction in errors.
- Doubled revenues from software renewals by analyzing usage metrics and designing new flow for product renewals.
- Introduced Scrum and led daily meetings of five engineers.
- Designed business plan for medical records product targeted at governmental organizations, growing revenue by 30%.
- Designed wireframes and built prototype for FDA approval product and successfully pitched execs on launching new division.

Senior Solutions Engineer (2003–2004)
Provided technical consulting, training, and ongoing support to industry-leading customers

- Consulting for deployment of ALM and CRM practices and software; achieved the highest level of customer satisfaction on every deployment (5 out of 5 customer rating).

Microsoft Consulting/San Francisco Unified School District (San Francisco, CA. 1999–2003)

Senior Database Architect

- Administered multi-platform network environment overseeing a five-person team of DBAs that supported 100+ database systems for all San Francisco schools.

EDUCATION

San Francisco State University

B.S., Computer Science

PROJECTS AND ACTIVITIES

- **Commercy (2012):** Built custom e-commerce website for local bike store, supporting shipping calculations, online store management, and custom themes. PHP, CSS, HTML.
- **SF Tutoring**, President (2010 - 2012): Ran tutoring program consisting of 100 low-income families and 40 tutors. Designed new tutor screening program, reducing annual turnover from 50% to 20% compared to prior president.

LANGUAGES AND TOOLS

- PHP, Visual Basic, C, JavaScript, SQL / MySQL, HTML, CSS.
- Balsamiq, Apache, AWS, Oracle, Marketo, Google Analytics, Photoshop, Fireworks, SalesForce.

Amit Agarwal (Anonymized)

Amit Agarwal received an internship offer from Google for an associate product manager role.

Original Resume

EDUCATION

Stanford University
Bachelor of Science, expected graduation in May 2013; GPA: 3.65, Projected major: Computer Science. Winner of Facebook Summer of Hack 2011

Mellow Brooks Charter High School
Diploma, June 2009, GPA: 4.0 / 4.0, SAT: 800 R, 800 M

Awards/Honors: Valedictorian, National Merit Finalist, National AP Scholar, McAllister

Scholarship, Alexis X. Wynn Scholarship

EXPERIENCE

Codesion
Software Engineering Intern, Summer 2011

Built Python system which continuously syncs bug database with any target system, such as search engines or integration platforms, while handling failure conditions.

Stanford Law School
Research Assistant, Spring 2010-Summer 2010

Assistant to Dr. Gilad Zalmon. Analyzed data from Michigan HRS to determine health effects of retirement. Researched publication bias in Corporate Governance articles.

Stanford Graduate School of Business
Research Assistant, Summer 2010

Worked with PhD candidate Michael Jameson and Dr. Patrick Revik to collect information on all publicly traded companies. Worked to build comprehensive industry classification standard.

ACTIVITIES

Stanford Upstart, President, February 2010 – present

Stanford ACM Team, Member, September 2010 – present (5th place at ACM regionals)

Stanford Club Soccer Team, Goalie, January 2009 – present

Stanford A Cappella, January 2010 - present

SKILLS

Computer: C, Python, Java, SQL, JavaScript (some), Ruby (some), Scheme (some)

Assessment

Amit actually has a fantastic background, but unfortunately this resume doesn't let his experience shine. In short, he's focused on the wrong stuff.

Like many (or most) other candidates, he focused too much on responsibilities and not enough on his accomplishments. His resume should be bulleted with three to five bullets about the ways he's had an impact on an organization. Bullets are your friend!

Additionally, it's interesting that his resume lacks projects. He is a student, so he's almost surely done some projects, at least for school if not for fun. It's very important to list these.

If we turn to his activities, we'll see a mix of things. Two of the activities aren't very interesting to us; he's unlikely to get a job where singing and soccer matter. One is *very* interesting, but it blends in. The "Upstart" activity might be interesting, but it's impossible to say because Amit hasn't said what it is.

We'll revamp his resume with these things in mind. We want to trim down the less important stuff (high school, sports, etc.) so his really amazing accomplishments stand out.

New and Improved

EDUCATION

Stanford University
B.S. in Computer Science (2009 - 2013). GPA: 3.65.

EXPERIENCE

Codesion
Software Engineering Intern (Summer 2011)
- Implemented syncing tool to synchronize bug database with other major bug databases to reduce switching costs for new users. (Python.)
- Extended tool to synchronize with bug reports posted on the largest online forums by using search APIs and parsing reports. (Python, Google Search API.)
- Designed new algorithm based on Alex Chen's Weighted Parsing algorithm to extract metadata (operating systems, milestones, etc.) from bug reports and optimized algorithm by 85%.

Stanford Law School
Research Assistant (Spring 2010 - Summer 2010)
- Proposed and built system to automate data analysis from Michigan HRS to determine health effects of retirement.
- Co-published paper in Advanced Maching Learning (AML) journal (Summer 2010) with Dr. Gilad Zalmon, who presented findings at AML's annual conference.

Stanford Graduate School of Business
Research Assistant, Summer 2010
- Collected information on publicly traded companies and helped with building comprehensive industry classification standard.
- Wrote Python script to extract data from online sources, reducing man-hours by 98%.

PROJECTS
- **EdU Projecto (iPhone App, Independent):** Project management app to help CS students working on school projects manage timelines and workflow. Downloaded 1000+ times in first three months with 4.7 / 5.0 average in iOS app store.
- **Billboard Processor (C++, Course Project):** Program which extracts text from photos of billboards. Achieved 95% accuracy, exceeding class average of 85%.

AWARDS AND ACTIVITIES

- **Facebook Summer of Hack (2011):** Placed 1st at weekend-long hackathon out of 47 teams.
- **ACM Regionals (2010):** Placed 5th out of 150+ teams on competitions involving data structures and algorithms.
- **Stanford Upstart**, President (February 2010 - May 2013): President of semi-annual hackathon with more than 100 students participating. Raised $10,000 in prizes from 30 sponsors, including Google, Microsoft, and Facebook.

ADDITIONAL INFORMATION

- **Programming Languages:** C, Python, Java, SQL, JavaScript (some), Ruby (some), Scheme (some).
- **Other Activities:** Stanford Club Soccer Team, Goalie (January 2009 - May 2013); Stanford A Cappella (January 2010 - May 2013).

Adam Kazwell

Adam received an offer from GigaOM with this resume.

SUMMARY

Early adopter with a talent for recognizing consumer value. Capable of creating a product vision and analyzing business performance. Excels at recognizing trends and opportunities in the consumer web industry, and passionate about building engaging products loved by millions of users.

AOL

Product Manager for AIM.com and Games.com (July 2011 - March 2012
Joined Jason Shellen's team to help with the re-launch of two key properties. For AIM, was focused on the web experience and helped to design the landing page and new user flows and also handled incoming user support across all platforms.

- Put together welcome experience for new and returning users of AIM to introduce them to completely updated product. Separately, used UserVoice to collect, aggregate, and respond to incoming feedback -- determined top addressable issues and shared summarized results with the team. Also began work on creating refined metrics dashboard.
- Before joining the AIM team, helped to determine refresh strategy for Games. com. Focused on tablet-driven design, the work was presented to senior leaders of the AOL consumer product group.

Hotwire

Product Manager (February 2009 - July 2011)
Translated new features into detailed business requirements and communicated their design and functionality throughout the organization. Worked across multiple verticals, doing full-sized project and ad-hoc work, using primarily waterfall SDLC.

- Worked with an in-house designer and 3rd-party team to create the first version of Hotwire's mobile-optimized site. Led decisions around which parts of the site should be included/removed and how options should be presented.
- Member of the launch team for Hotwire's international site, key strategic project in 2011. Helped to quickly establish a modular platform that provided core functionality while also being easy to iterate on.
- As part of cost savings exercise, designed templates and process flows for delivering an automated and extensible solution for using Amazon's Mechanical Turk for routine data collection tasks.

LiveJournal
Product Manager (April 2008 - January 2009)
Responsible for managing all parts of the product lifecycle, from ideation to launch. Wrote concept docs and PRDs, created initial mock-ups, and gathered metrics before and after product launch. Worked with developers in both the US and Russia to deliver key projects that helped to increase engagement.
- Reversed declining paid subscriber trend with redesign of account level management - making it easier for users to understand what features they would be getting at each account level.
- Used traffic metrics to determine unique product positioning opportunities and used insights to attract additional advertisers. Researched competitive landscape and market opportunities while communicating insights on product trends to senior management via PowerPoint presentations.

Yahoo!
Business Analyst (November 2004 - February 2008)
Provided reports and analysis on the overall performance of the Personals product to the executive team on daily, weekly, and monthly basis. Worked closely with the product team to define and track the impact of product releases.
- Created and maintained daily and weekly dashboards that tracked the seasonal trends, week-over-week, and year-over-year performance of over 100 product and referral metrics.
- Combined knowledge of key product drivers with experience as user of the Personals site to present insights to the product team that led to low-cost, high-impact feature development.
- Coordinated with Product team to define, track, and analyze benefits of each release. Decided which metrics would most likely be impacted in release and calculated baselines to confirm that their performance stayed within expected ranges.

TXU Energy
Marketing Analyst (December 2001 - September 2004)
Started as technical writer and transitioned into an analyst role for the Marketing and Business Information group. Responsible for gathering reporting requirements and turning them into technical specifications.
- Held requirements gathering meetings with multiple directors and drove the creation of standardized metrics and repeatable marketing performance reports, which were used across multiple lines of business.
- Used combination of SQL, Crystal Reports, and MS Access to pull data, which was then cleansed and formatted in Excel by parsing and concatenating fields, and summarized in pivot tables and charts. Created macros to speed up any repeatable processes.

EDUCATION

Marquette University, August 1997 - June 2001 (Milwaukee, WI)
Bachelor of Science, Major in Industrial Engineering

ELSEWHERE

Active throughout the consumer web. Most active on Twitter (@kaz), Quora, and Instagram.

Assessment

Adam's resume did fit on one page, but it was still very lengthy. Numerous blocks of text means his best stuff gets skipped over in just a brief glance.

Additionally, his resume lacked specifics and clarity.

For instance, consider his work experience at AOL. There was something about a "welcome experience," but it was unclear what that meant exactly. After that, he said he used UserVoice to collect feedback, shared results, began work on something, and helped with something else. None of these are really clear accomplishments.

His description of his work with HotWire has similar issues. He "worked with" some people, "helped to" do something else, was a "member of" another group, and participated in some "exercise." These lack the "umph" they could have.

With LiveJournal, we see some clear accomplishments, but they could still be better. The part about "reversed declining paid subscriber trend" is fantastic, but it would be even better backed up with numbers. He says he used traffic metrics to determine positioning and attract advertisers. "Attracting advertisers" is just slipped into the middle of the bullet and isn't backed up with numbers; even if a reader noticed it, it wouldn't be very credible.

Adam's resume was by no means terrible; in fact, it was better than most resumes. It still could be so much better, though.

Adam took this feedback and redid (and updated) his resume. Here's what he came up with. Observe how:

- His AOL work is clearer (we can now understand what the welcome experience was) and more focused on his own accomplishments.

- His HotWire work shows more leadership. It's now about what he led, not who he worked with. He also now has numbers to back up how he reduced costs. We also now learn it wasn't just an "exercise." This was something performed with a meaningful impact.

- For LiveJournal, he's provided numbers on how the subscriber trend changed. That makes them more credible *and* makes someone more likely to notice that in a quick skim. He's now stated he attracted two new advertisers (although this might be even stronger if he could say how much money that brought in). He's also added a new bullet about growing connections.

- His summary is more meaningful. Anyone can claim they have a "talent for recognizing consumer value," so that's not very meaningful. The new summary feels more tangible.

- He's kept his social networking usernames in. This can be a nice way for companies to learn more about you.

His bullets are still a bit lengthy, but his resume has improved substantially. Before, we would walk away from his resume with a general impression of what sorts of things he was responsible for. Now, we know that not only was he responsible for some interesting things, but he's also been successful. That's what you should be trying to do in your resume.

New and Improved

Consumer web product manager with 5+ years experience. Engineering background. Worked as a business analyst before transitioning to a product manager. Focused on finding the right questions to ask and assembling and iterating on the best ideas to build great products.

EXPERIENCE

GigaOM
Product Manager
Responsible for the main parts of the Gigaom.com WordPress-powered blog, including the front page and story pages. Led responsive site redesign. Reported on and analyzed key product metrics via Google Analytics and Chartbeat. Drove agile development via Asana and Github.
- Product lead for the GigaOM.com responsive redesign. Set overall direction for what was then Gigaom's largest project to date. Post launch, mobile traffic time-on-site doubled, and visitor traffic at non-peak times grew by 20%.
- Took over a "Analyst Connect" product that had been stagnant for months and created a clickable mock in Powerpoint. This allowed for improved iteration amongst the team and led to a polished product getting shipped one month later.
- Adapted WordPress's post formats to deliver new customized post types. This

allowed authors to increase their output by 25%, leading to record unique visitor and pageview levels.

AOL (AIM.com & Games.com).

Product Manager (July 2012 - Aug 2013)

Joined Jason Shellen's team to help with the re-launch of two key properties. For AIM, was focused on the web experience and helped design the landing page and new user flows. Also handled incoming user support across all platforms.

- Outlined and designed welcome experience for new and returning users of AIM, introducing them to a completely re-imagined product - included landing page layout and messaging and feature walkthrough post login.
- Managed UserVoice account to collect, aggregate, and respond to feedback from millions of users. Determined top addressable issues by analyzing stats and led team to resolve these issues.
- Before and while working on the AIM team, helped create and communicate the refresh strategy for Games.com - including getting buy-in from key executives. Reshaped roadmap to address new goals.

Hotwire

Senior Functional Designer (Feb 2009 - Jul 2011)

Took rough feature requests and turned them into detailed business requirements. Determined the specifics of a feature's design and functionality, and communicated their final planned state throughout the organization. Worked across multiple product verticals, doing scheduled project and ad-hoc work, using primarily waterfall SDLC.

- Guided the first version of Hotwire's mobile-optimized site with a single in-house designer and third-party development team. Led decisions around which parts of the site should be included/removed and how flow should be optimized.
- As a member of the launch team for Hotwire's international site, defined plan to build a modular platform to accelerate iterations so we could quickly adapt after entering a brand new market.
- Reduced costs for routine data collection tasks by 90% by creating templates and process flows for an automated and extensible solution using Amazon's Mechanical Turk.

LiveJournal

Product Manager (Apr 2008 - Jan 2009)

Responsible for managing all parts of the product lifecycle, from ideation to launch. Wrote concept docs and PRDs, created initial mockups, and gathered metrics before and after product launch. Worked with developers in both the U.S. and Russia to deliver key projects that helped increase engagement.

- Reversed a year-long paid subscriber decline from -3% month-to-month to +2% month-to-month with redesign of user account level management, making it easier for users to understand which features they would get at each account level.

- Navigated complicated privacy settings and designed a plan to update "Find a Friend" functionality, growing new user connections within first 90 days by 10%.
- Analyzed traffic metrics to discover and promote a collection of journals that attracted two new advertisers.

Yahoo!
Business Analyst (Nov 2004 - Feb 2008)
Provided the executive team with daily, weekly, monthly, and quarterly reports and analysis on the overall performance of the Personals product. Worked closely with the product team to define and track the impact of product releases.

- Created and maintained daily and weekly dashboards tracking the seasonal trends, week-over-week, and year-over-year performance of more than 100 product and referral metrics.
- Led product team on selecting metrics to define, track, and analyze the benefits of each release. Decided which metrics would most likely be impacted in a release and calculated baselines to confirm their performance stayed within acceptable ranges.
- Drove development of low-cost, high-impact features by presenting insights on key product drivers to product team.

EDUCATION

Marquette University (Milwaukee, Wisconsin)
Bachelor of Science in Industrial Engineering

ELSEWHERE

- Active throughout the consumer web.
- Specifically on Twitter (@kaz), Quora, and Instagram (@kaz)
- More: about.me/kaz

Cover Letters
Chapter 9

Cover letters are a mixed bag. Some companies care about them a lot, some companies don't want them at all, and a lot of companies ask for them but don't put much emphasis on them.

A well-written cover letter can help connect your background with the job requirements. This becomes particularly important if you don't have the "ideal" background. It can be your way to show the reader that, although it doesn't look like it at first glance, you have the right background.

Elements of a Good PM Cover Letter

Most cover letters are mediocre. They merely restate the candidate's resume in paragraph form. This might show some basic writing skills, but it doesn't add (or subtract) much.

A good PM cover letter has the following attributes:

- **Short:** Keep your cover letters to around 200 - 250 words. Lengthy cover letters are less likely to be read and likely contain a lot of unnecessary detail. Additionally, being too verbose will reflect poorly on you.

- **Shows Passion:** An ideal PM is passionate about technology and about the industry. This passion should be reflected in the cover letter. Tell the reader why this job excites you.

- **Demonstrates Skills:** Look through the desired skillset of a PM – in general and for this position – and reflect on how your background does or doesn't match it. Pay particular attention to any skills you possess but which might not easily be shown in your resume; use your cover letter to demonstrate these.

- **Matches the Company Culture:** Does the company prize "business people," or does it like to show a bit more flair? Reflect this in your cover letter. A fun, quirky company could merit a fun, quirky cover letter – particularly if your resume otherwise reads like a boring business person's.

- **Well Written:** Your cover letter is a writing sample and it should be handled as such. This means, of course, no spelling or grammar errors. It should also be succinct rather than full of flowery, descriptive prose. Keep an eye on your sentences and be sure they don't get too long or complex. Vary between longer sentences and shorter sentences.

Ultimately, your cover letter is a way to connect your background with the "perfect" PM fit. Use it to do that.

The Cover Letter Template

A traditional cover letter fits a fairly standard template. The exact order of these paragraphs (namely the second and third) can be rearranged, but a good cover letter typically includes these elements.

Addressing the Reader

A simple "Dear _____" works fine. If you know the name of the person you're addressing, use it. Never, ever say "Dear Sir" unless you know for a fact the reader is male. That is an excellent way to get rejected.

> *Dear _____,*

Opening Paragraph

In the first paragraph, briefly introduce who you are and what position you're seeking. There's no need to state your name; that will be in your signature. If you have a personal connection with the company or an interesting way you heard about the position, this is a good place to mention it.

> *I recently attended the WYSIWYG Developer Conference, and I was thrilled when ChattyCha's CEO mentioned an opening for a product manager on the API team. As a developer-turned-PM, I believe this position would be an excellent fit for my background and interests.*

Second Paragraph

Here you discuss how your background makes you a good fit for the position. This should not be a summary of your roles; that's what your resume is for. Rather, what this paragraph does is connect your skills and accomplishments with what the company is seeking.

This paragraph should highlight your soft skills and back them up with accomplishments.

> *I am a deeply technical person who loves motivating others to achieve a difficult goal. During my first year as a developer at Microsoft, I found a way to resolve a long-standing issue with a key framework by leveraging a new technology. I recruited and trained a team of three engineers, and later became dev lead on a related feature. I took this passion for problem solving to Waffle, an enterprise-security startup, where I successfully launched the company's flagship feature – one that many thought would fail due to its complexity.*

Third Paragraph

Next, you explain why you're excited about this role. You don't want to make this paragraph too long, as the prior paragraph is more important. You just want to show you care about this position and it's not just another company you're applying to.

> *I am excited about the challenge of ChattyCha's mission. I love deep technical challenges. The developer tools space is particularly interesting to me, having experienced the frustration of inadequate tools for my personal and professional coding projects. I am eager to help make this space better.*

Fourth Paragraph

The final paragraph is very short and just concludes the letter with a "thank you."

> *Thank you for your consideration. I look forward to hearing from you about this opportunity.*
>
> *Sincerely,*
>
> *Your Name*

A Great Cover Letter

Following the above template won't win you any awards, but it will basically get the job done. The job, of course, is to get someone to consider your resume (or at least not rule it out).

On rare occasions, a cover letter can substantially increase a candidate's odds.

Consider this cover letter for a candidate at an education startup:

> As you might notice from my 2.7 GPA, academics have never really been my thing. But I love learning, and it's the delta between academics and learning that makes me so excited about FusionEd.
>
> I was only a so-so student. School's grades-driven approach didn't allow me to spend less time on the subjects I hated in favor of those I loved. And, even if I wanted to explore a topic more deeply, I lacked the resources and structure to do so. FusionEd's mission of child-driven education will tackle this issue and I want to be a part of that.
>
> Academics aside, I truly love learning, and especially learning about technology. Last year, I taught myself to code and I've since entered several hackathons. In the latest one, I was proud to be the only single-person team to receive an award. You can see some of my projects on my GitHub profile, including one targeted at students.
>
> In my professional life as a product manager with Colapa, I lead a team of six developers. My day-to-day work includes everything from feature design, to market research, to simple coding. I wear many hats, and it's one of the things I love about the role. One of my proudest accomplishments though was finding a way to reposition our product to capture the enterprise market. This now represents 15% of revenue.
>
> Most recently, I worked on developing an API for Colapa. I understand that FusionEd is heading in a similar direction, and I am excited to take on this challenge again. I am confident my love for technology, learning, and leadership – plus my dislike for the status quo in education – will enable me to make an impact on FusionEd.

This wouldn't necessarily be a great cover letter for all companies, but it's a great cover letter for the right company. It's honest and sincere, with just a little bit of a quirky punch in the beginning. The tone will work beautifully for a smaller startup that cares deeply about its mission to transform education.

On the skills side, this candidate has shown:

- Initiative (by learning to code).

- Passion for technology (by learning to code and by competing in hackathons).

- A willingness to take risks (by competing in hackathons, despite being vastly under-qualified).

- Intelligence (by winning hackathons).

- Leadership (by repositioning the current product).

- A love for entrepreneurship (by being at a startup and by "wearing many hats").

- Successfulness (by successfully repositioning the product).

This is the way you should dissect your cover letter. Think carefully about the role each sentence plays. What skills and attributes are demonstrated in your cover letter? Remember that your cover letter does not need to be a comprehensive list of all your experience; your resume is attached, too.

Company Research
Chapter 10

Knowing the ins and outs of a company can impress your interviewer. You'll be able to ask more interesting questions, give more insightful answers, and show more excitement for the role.

Additionally, sometimes interviewers will forget that not everyone outside the company knows the products as well as they do. They might judge you harshly for being unaware of some seemingly "obvious" details, or you might do poorly on a question because you didn't have some bit of background knowledge.

This section will outline what you should know about the company.

The Product

You should understand what the company is doing at a deep level. A cursory "it makes home automation tools" level of understanding is not sufficient. Knowing the following information will help you stand out:

- **Products:** What is the array of products or features that the company creates? How do the products fit with each other?

- **Competitors:** Who are the competitors? How does the company differentiate itself from them?

- **Customers/Market:** What is the target market for the company? Are there any secondary markets right now or that you would suggest the company enter?

- **Revenue:** How does the company make money? How would you suggest the company make more money? If it doesn't make money, what revenue strategies would you want to explore?

- **Love and Hate:** How do customers feel about the product? What do they love or hate? What are the most common complaints and issues?

- **Metrics:** If possible, try to learn about the company's key metrics. Finding exact numbers might be difficult, but you can at least get an understanding of which metrics they're doing well on and which ones they're struggling with. How many users does it have? What is its conversion rate? What is its growth rate?

- **News and Rumors:** Have there been any interesting news reports about the company? What is the company rumored to do? Don't just read this stuff. Formulate an opinion on it.

To learn this information, check out the company website, their About Us pages, their blog, their job sites, their SEC filings, newspaper articles, blog articles, their support pages, and whatever pops up in online searches.

You should also *use* the product yourself, and you should use it extensively across multiple user types. If it's a product with free and paying users, try to use it in both scenarios (if possible). Think about what parts you enjoyed and what you didn't. Pay particularly close attention to anything that would have turned you away if you weren't "forced" to check it out.

The Strategy

You should know not only *what* the company is doing, but *why* it is doing it. Knowing the "why" will help your answers fit the company's view of the world. For example, if the company cares about "changing the world," then you should mix that sort of passion into your answers. Or if the company is passionate about expanding worldwide, you can talk about global expansion and internationalization issues.

Knowing the "why" means understanding the following:

- **Mission:** Look up the company's mission statement. How does it live up to this mission? Be specific.

- **Strategy:** What do you think is the company's strategy? Are there any missteps with respect to that?

- **Strengths:** What are the product's selling points? How does the company leverage those? What about the company or its products has enabled its success?

- **Weaknesses:** What are the major issues with the company and its products? How should the company address those weaknesses, or should they just accept them?

- **Challenges:** What are the biggest challenges for the company right now? How do you see them addressing those? What challenges have they overcome?

- **Opportunities:** Is there anything on the horizon (with technology or in their industry) that might create an opportunity for the company?

- **Threats:** Similarly, is there anything on the horizon which might threaten the company's success?

- **Future:** What do you think the future holds for this company? Think about any new products or features that would be a natural fit.

These topics are just to get you started. Your responses to the above questions might well overlap, and that's okay. If you can make a compelling argument for why the product will (or won't) succeed and predict arguments to the contrary, you're probably in good shape.

The Culture

Some candidates get so focused on the product that they forget it's the people who make the company (which in turn makes the product). Or they do some quick "online stalking" of their interviewers but don't think about the aspects that really matter.

- **Culture:** What is the company's culture? The company's job pages might discuss this a little, but that's obviously going to be biased toward the image they want to project. Look online for reports of what candidates, current employees, and former employees say about working there.

- **Values:** What does the company value? By "values," we mean anything that's important to them, explicitly or implicitly. To understand this, read interviews with the founders and think about their culture and products. Values might include aspects such as moving fast (Facebook) or "don't be evil" (Google).

- **History:** How long has the company been around? How did the company get started? Have they been doing what they set out to do, or have they pivoted?

- **Interviewers:** If you know who will be interviewing you, you can search for more information about them online. Don't be creepy, though. Congratulating an interviewer on her recent marriage comes off "stalker-ish"; it's best just not to mention things like that.

- **Key People:** Who founded the company? What were they doing previously? If it's a startup, who funded the company? Are there well-known people at the company? These aren't just "facts" to know; you should think about them. How does the background of the founders and other key people affect the company?

- **Organization:** How big is the company? How is the company organized? Does everyone basically report to the CEO or is there a strict hierarchy?

The point of this knowledge is not to whip it out to impress the interviewer (it probably won't), but to use this information to get a feel for what the company is really like. You want to make sure a company is a good fit for you, too. Sometimes candidates are so focused on landing the job that they forget to make sure the job is really right for them.

This information has another benefit, too: it can help shape your responses to behavioral questions, like how you influence teammates or how you would implement a decision.

If you can, try to fit all the little learnings together. That is, match the culture to the strategy and other parts of the company. For example, Amazon has a reputation for being very money and metric driven, as well as very frugal. Is this because Amazon is a relatively low-margin business? Is it because Amazon was not profitable for a long time? How does this affect its strategy?

The Role

Finally, you should know how you fit into the company. This entails the following:

- **The Role:** You should understand the role of a product manager at this company. How technical is the role? How do decisions get made?

- **Idea Generation:** Where do ideas come from? Some companies are "bottom up," with ideas coming from developers and product managers who then convince the executives of their vision. Other companies are "top down," with executives supplying long-term vision for the company and PMs being tasked with implementing that vision.

- **Practical vs. Crazy:** While all companies have some appreciation for the practical realities and some for the crazy off-the-wall ideas, each firm strikes its own balance between the two. Understanding if the company loves bold ideas or prefers more incremental improvements will help you understand what sorts of ideas you could pitch.

- **Things to Change:** Walk in with some ideas for what you'd want to change or implement at the company. An understanding of major user complaints will give you a good place to start.

- **Why You Want the Job:** You should be able to speak passionately about why you are excited about the company *and* why are you excited about product management.

- **Why You Would Be a Good Fit:** Similarly, you should be able to make a compelling pitch for how the role matches your skillset and background. The job description is a good place to start, but think beyond this. What challenges does the company face and how well suited are you to tackle them?

An interview is a good place to get a deeper understanding of some of these aspects, but you shouldn't rely on this. You'll want to walk in with as good an understanding as possible.

The Questions

With all the research behind you, it's wise to walk into an interview prepped with questions to ask the interviewer. Almost every interviewer will give you an opportunity to ask them questions, and you don't want to enter that "awkward silence" phase.

You can think about questions as falling into a few major categories:

- **Useful Questions:** The "Do you have any questions for me?" moment is, of course, a useful time to get some things answered about the role or company. It can be particularly valuable to ask a question where the answer might vary across interviewers. For example, "What do you find most challenging about being a PM here?"

- **Passion Questions:** Some questions show a passion for the company or role, simply because you had the background or desire to ask the question. For instance, asking a company why they dropped their initial monetization strategy shows you've researched the company, and asking a question such as, "Where do you see the company going in five years?" shows the drive to understand the company.

- **Expertise Questions:** Other questions do more than just show research; they can show a deep understand of the company's business. For example, the following question shows an understanding of international markets: "If it's something you can share, I'd love to learn more about why your company decided to enter Europe before Asia. Given the regulatory issues in Europe and the smaller market size, Asia would seem like a more natural pick."

In addition to these general types of questions, some questions work well in many interviews. Some good questions include:

- What's a typical day like for you? How much of your day do you spend writing specs versus working with designers versus doing other activities?

- How has the role of a PM changed? How do you see it changing?

- What is the balance between PMs, developers, and designers? How does decision making work?

- What's your favorite part about working here?

- What would make someone the ideal PM candidate for you?

- What do you think makes this company's culture unique?

- What do you find most challenging about being a PM here?

- Who do you work with on your core team versus extended team?

Don't underestimate the importance of questions for your interviewer. Many candidates find their interviewer leaves a ton of time for questions, and they just don't have enough questions to ask the interviewer.

What Not to Ask

Do not ask questions that are:

- **Red flags:** For example, asking about vacation time in an interview can come off the wrong way. Just how much vacation are you planning on taking?

- **Obvious:** Asking a question with obvious answers suggests you haven't done your homework and you may not be very serious about the role. For example, a question such as, "What's the monetization strategy?" might, in many cases, reveal a lack of preparation.

- **Critical:** A question like "Why didn't you do _____?" can sometimes come off as negative, or as an I-want-to-prove-I'm-smart question. Such questions can also just be annoying, particularly when the answer is likely to be some version of "too many features, too little time."

If you're worried a question might reflect poorly on you, you can always ask it after you have an offer.

Define Yourself
Chapter 11

There is a huge range of potential questions, but these six questions come up so often that it's worth spending some special time on them. For each of these questions, you should have an answer prepared. This means not only knowing what you'll say, but actually rehearsing the answers.

"Tell Me About Yourself" (The Pitch)

Many interviewers will kick off the session with an open-ended "Tell me about yourself." question. In fact, over the course of a full day of interviews, you can almost guarantee you'll get this question. You should be prepared with a solid pitch about your background, accomplishments, and interests.

This is not a time to just read off your resume or blab about your personal life. Rather, pick out a few of the key things you'd like your interviewer to know about you. This is an opportunity to connect your experience with the job you're interviewing for.

Consider this pitch:

Interviewer: Tell me a bit about yourself.

Candidate:

> Sure, I'd love to.
>
> I'm currently a PM with MapSign, where I have been for the last two

years.

I got into product management by accident, but found that I loved working with customers. A lot of people see technology as forcing complexity into their lives, but I don't think it has to be that way. I like listening to customers, understanding exactly what their issues are, and figuring out how technology can simplify their lives.

After I graduated from college, I joined a small startup called MealRight as a software engineer. We were doing okay, but it just didn't look like we were getting that hockey stick growth that everyone talks about.

I knew a bit about the restaurant industry from my family's restaurant growing up, and I proposed pivoting to focus on giving restaurants more visibility into their suppliers. I talked with a bunch of restaurants to validate this problem, and successfully convinced the team to shift gears.

I stepped in as product manager where I stayed for several years, eventually leading a team of four PMs.

I'm currently a PM with MapSign, where I manage the personalization features.

I recently launched a feature which allows users to create their own maps. This was actually created on a whim for a hackathon I participated in, but it turned out that users really loved this. It allowed them to tell a "story." Whereas previously people used our product mostly privately, now people were sharing their maps with family and friends. This has resulted in about a 15% increase in paying users.

I can go into more detail about my work at MapSign if you'd like.

Outside of work, I do a bit of coding for fun, participating in hackathons like the one I just mentioned. I also run a blog focused on the New York City music scene. I manage a team of ten writers—mostly college students, who are happy to write for free. We get about 200,000 visits per month.

I'm looking now for a role at an early-stage company where I'll be able to take a single product from conception through to launch. I think my experience as a developer and as a PM who works cross-functionally will give me the background to work with, and even take on some of the responsibilities of, several other roles. That's a lot of what interested me in this role; it's at the perfect time for me to dive in, and it's so closely related to the personalization work I'm doing right now.

In this pitch, the candidate has walked the interviewer through his background, while sprinkling the story with key accomplishments, connecting the different accomplishments together, and prompting the interviewer to ask more about the areas where he shined. He's formed a cohesive story about what he's good at and why he loves being a PM.

Design your pitch by thinking about what you want the interviewer to know about your background, experiences, and interests. Where possible, connect elements of your pitch to what the company is looking for, whether that's aspects of the PM role specifically or the company's product.

Additionally, keep in mind these dos and don'ts.

- **Do** be mindful of how long you speak. Your interviewer will be judging you in part on your communication skills, and no one likes a rambler. A good rule of thumb is to speak one to two times as long as your interviewer did, if she introduced herself. Otherwise, about two minutes is a good guide.

- **Do** highlight the most interesting or relevant parts of your jobs. This is your opportunity to sell yourself.

- **Don't** just list off your accomplishments. That can come off as too boastful, and your interviewer might even get lost in all the details. Your pitch should be a cohesive story about how you got from then to now. It should connect the different elements of your life and offer context for why you're a good fit for this role.

- **Don't** get overly technical. It's great if you have some very technical experience, but your interviewer wants to hear in more straightforward language why that work was important. After all, as a PM, you need to communicate with both technical and non-technical people.

- **Do** discuss your "extracurricular" activities, where relevant. For example, if you're applying to a fitness-related startup and you've started a marathon training group, that's a good thing to mention. It shows a passion for the space, and possibly even expertise in it. Even when the extracurriculars are not directly applicable, they can often show initiative and leadership.

- **Do** practice. It's almost a guarantee you'll be asked for your "pitch," so it's silly to be unprepared. Grab a friend and let them hear your pitch. How does it sound to them? You might even want to record your pitch for yourself so you can hear how it sounds.

- **Don't** speak too abstractly. Instead of just saying you did customer research and wrote specs, talk about an example of something important you learned and how you changed the product design based on that.

- **Don't** be boring and just rattle off a bunch of facts about yourself. Weave your pitch into a mini-story.

- **Do** be passionate and proud of your past work.

You should look forward to this question. This is your opportunity to sell your interviewer on why you're the perfect candidate, and you'll be perfectly prepared to do so.

"Why do you want to work here?"

When I interviewed candidates at Google, I would ask them why they were interested in the position. Many answers would be something along the lines of:

- Because I've heard Google has a great culture.

- Because Google has changed the world.

- Because so many people use Google products, and I want to work on something my family and friends have used.

These answers are okay, but they're just okay. You won't win any points delivering an answer like that. Why? Because you haven't given me anything that makes me want to hire you more.

An ideal answer will sell yourself in some way. Consider integrating one or more of these aspects into your answer:

- **Company Research:** You can use your answer to show that you've done research about the company or position. Doing research shows passion for the position, and passionate employees make good employees. For example, you might say, "What piqued my interest was a presentation your lead UI designer made about the different methods you use for data analysis, and how this helps you build a better product. I'm very quantitative, so I really want to work for such a data-driven company."

- **Relevant Experience:** Your answer can actually communicate to the interviewer directly that you have relevant skills or experience. For example, you could say something like this: "I'm really interested in testing tools. For my senior project in college, I built an automated way of detecting certain types of errors. I learned a ton about different types of website errors and ways to detect them. I was really intrigued by how much impact automated testing can do, if done well—but I also learned just how challenging it is to do well.

I'm excited to get back into this space and leverage what I learned."

- **Passion:** It can be valuable to directly communicate passion for a position. This is especially true for startups that are focused on some sort of "greater good" for the world. For example, for an education-related startup, it could be good to have an explanation like this: "I grew up fairly poor and attended pretty mediocre elementary schools. What helped me excel beyond my peers and win a scholarship to a top high school and college was seeking out extra resources. At that time, this meant public libraries, mentors, and a small number of websites. I think your company has the potential to bring these online resources to a much broader community of disadvantaged students. This was so meaningful to me growing up that I really want to be a part of that now."

In general, big companies will be less impressed by your having researched the company or being really passionate about working for them. They presume, for better or worse, that you want to work for them. It's better to show a passion for the role/team, or some experience that makes you a good fit. Avoid answers about it being a good stepping stone for your career or a nice name for a resume. Those answers might make your interviewer think you're not going to stick around for long.

At a startup, passion for the company or expertise in the space can be really important. You should use the company's product and have an opinion about it. Avoid answers about wanting the "financial upside" of a startup. That doesn't make you a better candidate; it just means you like money. It also means you might be unrealistic about the hardships of a startup or that you might leave the second things get rough.

Try to come up with two or three reasons prior to your interview. And, of course, practice your answer.

If you're asked this question by multiple interviewers, it's okay to give a very similar answer to each. In fact, it might be strange if you didn't offer similar reasons.

"Why should we hire you?"

This is an aggressive phrasing for the question, but it can be asked in many ways. An interviewer might instead ask you why you think you would be a good fit for the role or what you think you could offer the company.

As intimidating as this question can be, it's actually a great question for you. You're being asked to sell yourself; no more subtly inserting little tidbits about why you'd be a great fit. What could be better?

Your answer to this question can include any or all of the following:

- **Why you are a good PM:** Have you shown initiative in your current job? Do you have deep technical skills? Have you been successful as a PM in the past? Where possible, back up your answer with evidence. It will be much more convincing.

- **Why you are a good fit for this space:** Have you worked in an area relevant to the team or company? Are you really passionate about the space? Why? This is a good time to mention what you know about the company or team's industry. For example, if you know a bunch about advertising, talk to the interviewer about how you have experience in this area.

- **Why you are a good fit for this company's culture or work environment:** Sometimes, there can be unique aspects of the company's environment that make you a good fit. For example, suppose it's a very small startup where PMs take on a lot of responsibilities beyond those traditionally belonging to a PM. You might say something like: "I enjoy taking on a diverse set of responsibilities. Outside of work, I started a volunteer organization tutoring children. This has given me the opportunity to juggle responsibilities like marketing, advertising, event planning, management, and recruiting. I've found I really enjoy doing a diverse set of tasks at once, and I'm not above taking on the tedious work."

The more you know about the company or position requirements the better. Before your interview, re-read the job description and come up with examples to match what the company is looking for.

Many of the elements of your answers to "Tell me about yourself." and "Why do you want to work here?" apply to this question, but you would use these elements more directly.

A good answer to this question might be something like:

> I think three things make me a great fit for this position.
>
> First of all, I have several years of experience as a PM, and I've consistently shown success in the role. I've launched four critical features from scratch in that time, and was rated the top PM at my company. This is what led to my current hiring manager recruiting me to his team.
>
> Second, as a CS major and statistics minor, I have the technical and quantitative background that you're looking for. I haven't done a ton of data analysis in my current role, but I have the academic background and raw skills to learn it quickly. I've actually started learning a bit of

data analysis through some online courses. I have no doubt I'll be able to pick up the skills you need quickly.

Third, I really want to be here. I'm a fitness junkie, having run three marathons and participated in two Tough Mudder races. I truly believe in your mission of the gamification of exercise and I'm constantly coming up with ways of making exercise more fun. I've even recruited several of my anti-exercise friends to regularly workout—and enjoy it.

This candidate has done several things in this answer that you can learn from. First, he's shown he wants to be there and he can do the job well. Second, he's offered information to compensate for what an interviewer might perceive as a gap in his expertise: data analysis. Third, he's used the opportunity to list specific, concrete accomplishments that wouldn't otherwise appear on his resume, such as being recruited by a manager. Fourth, he's tackled the answer with structure.

Responding in a well-structured way to a fairly open-ended question will not only demonstrate to your interviewer that you are a strong communicator, but it will also help your interviewer retain the information you give him.

"Why are you leaving your current job?"

Often, this question is asked as an icebreaker. Your interviewer isn't necessarily looking for anything specific; she just wants to learn a bit more about you. Your goal is essentially to not screw things up.

What can screw up an answer? Any of these things:

- **Complaining about your current role.** Interviewers don't like negative candidates. They worry that candidates will always be negative—even after being hired—or that the candidate's dislike of her current role is actually her own fault. If you have a bad relationship with your boss, this could be as much your fault as it is your boss' fault.

- **Focusing on money.** A candidate who is just looking for more money often doesn't make a great employee. An interviewer might worry that such a candidate lacks passion and commitment for the job, and will leave that company as soon as the opportunity for more cash comes up.

- **Being bored and not learning.** While there are ways to spin this in a positive way, being bored at your current role means your most recent experience isn't very valuable or interesting. This is probably not the way you want to kick off your interview.

Instead, focus on the positive. What are the things you are looking forward to in a new role?

Any of these reasons, which range from benign to positive, would work as an explanation:

- "To be honest, I'm actually pretty happy in my current role and wasn't really looking to leave. I stumbled across this job opening though, and it sounded like such a great opportunity I just couldn't pass it up."

- "After being a development manager for several years, I've acquired deep technology and leadership experience. I now want to take those skills to move more into the product management side, since I really love the experience of launching a new product."

- "I've had to relocate for personal reasons."

- "My company was acquired last year, so it's changed a lot. I'd like to be back in a more entrepreneurial environment."

- "I actually really like my current company and role, but after being there for several years, I want to be somewhere which is giving more back to the world."

- "I have a really deep statistics background and I really enjoy being quantitative. I want to be somewhere where I can use this more directly."

- "The current team I'm on is more enterprise focused, and I want to move closer to the consumer."

- "I've gotten a lot of great opportunities in my current role, but after being there for three years, I feel my learning has slowed down a bit. The company and space just aren't growing that much. I want to take on new challenges."

This is not a time to blabber for a while. Keep it short and sweet; one to three sentences should be enough to handle this question.

"What do you like to do in your spare time?"

With this question, your interviewer is primarily looking to get to know you and what you're interested in. The best answers to this question might show some sort of experience that's relevant to the position, but discussing any passions that you have outside of work will be a positive.

Best Answers

The answers below show some relevance to the job by highlighting the candidate's leadership abilities, technical talent, or initiative. All of these things make the candidate well suited to a PM role.

- "I really enjoy exploring new technology. I've set a goal to build a simple to-do list application in as many languages as possible, just to learn a bit about what each programming language does. So far I've done this in five languages."

- "I've been getting involved recently with online learning. I've taken a few online courses to learn a little bit about various roles, such as marketing, advertising, and finance. I'm also preparing to launch my own class on Product Management for Software Engineers."

- "I've been doing a lot of volunteering for the animal shelter. A few months ago I created and launched a new program called Teens 4 Animals to get more teenagers involved. That program has recruited 50 new volunteers from local high schools and saved more than 100 dogs from being put to sleep."

Good Answers

The answers below don't show any particular relevance to a PM role, but they do show some interests outside of work. Most importantly, they're all backed up with some concrete evidence that makes the interest sound more convincing.

- "I read a lot. Most recently I've been reading a lot of psychology books. I'm really fascinated by how our minds work and why we make the choices we do. So much of it is counter-intuitive."

- "I love playing ultimate frisbee. I play on a frisbee league at my company, and also joined a neighborhood pickup league."

- "I really enjoy doing basic construction projects around my house. I still have a lot to learn, but I've had some recent successes. I re-did my patio a few months ago and it's still standing. I guess that shows some promise."

- "I'm the lead singer and primary songwriter for a punk rock band called Out Like Pluto. I co-founded the band with one of my bandmates, and I also handle most of the social media marketing for the group."

No-Value-Add Answers

It's difficult to really screw up this question, but there are plenty of answers that won't really help your case. The answers below are weak because they don't show any passion. You go to work, come home, and then see your family and friends. This answer won't hurt you a ton, but you're losing the opportunity to add a new dimension to your job.

- "I'm not sure. I guess I don't really have a ton of true hobbies since I work a lot."

- "I mostly just hang out with friends. We go out to dinner, sometimes wine tasting or something like that. I've also organized the occasional barbeque at my place."

- "I spend a bunch of time just reading things on the internet—blogs, tech news, etc."

Most people do have some hobbies and that can be backed up with evidence, if they think hard enough. Do you work out daily? Set a goal for yourself, like training for a marathon or developing ten new exercise routines per month. Do you spend a lot of time on the internet? Use that time productively by writing a blog or answering questions on Q&A forums. Do you enjoy photography? Take some classes in it or set up an online portfolio, even if it's not very good.

You are not trying to account for all of your free time in this answer. You're just trying to add a new dimension to your profile and talk about it credibly.

"Where do you see yourself in five years?"

This question scares candidates, in part because they think that interviewers are asking it to trip them up in some way. Not so.

An interviewer asks this question for several reasons:

- **To understand if you actually want this job:** If the role doesn't fit into your long-term goals, it's a good tip-off to your interviewer you are just looking at this as a short-term job to fill in the gaps. Moreover, if you're not prepared to answer this question, it might be a sign you don't really care about the job.

- **To see if you even have a plan:** Successful people tend to know where they're going in life. If you don't have a plan, an interviewer might worry you're not very serious about your career.

- **To test your ambition:** Ambitious people generally make good employees. If your long-term goals reflect a lack of ambition, that's not a good sign. Similarly, if you're unrealistically ambitious, that also isn't a good sign. You have expectations the company can't match, and you're unlikely to be a good fit.

- **To ensure the company can give you what you want:** If a role isn't a good fit for your long-term goals, the company wants to know that. You'd be unhappy in the job, would likely be unsuccessful, and would quickly leave. That's not a good thing for anyone.

Ideally, you will have 5-year, 10-year, and 20-year plans, and this role will fit nicely in line with that.

If you don't have things quite as nicely planned as this, you can think of the answer to this question a different way: Where would very strong performance in this role get you in five years? Hopefully, that is where you see yourself.

If you're prepared for it, this question can be a great opportunity. You can use it to tell the interviewer whatever you want about yourself.

> I'd love to be lead PM for an emerging business unit where I'll get to think about long-term strategy, particularly with respect to monetization models. I'm particularly interested in mobile, and I could see myself leading a team of PMs and engineers on the mobile side. Additionally, I'm really excited about the ways in which a company can help new employees ramp up faster. I could see myself creating a more formal new employee training program where employees get mentorship and training in different areas of the business.

With this answer, the candidate has demonstrated ambition as well as a passion for business models, leadership, and mentorship.

You can use your answers to the career vision question to highlight what you're good at and what most excites you.

"What are your strengths and weaknesses?"

While many interviewers have moved away from this question, it's still common-place enough that you should be prepared for it. Be prepared for your interview with three strengths and three weaknesses.

For your strengths, focus on things that are specific, relevant to the position, and can be supported with evidence. For example, you could say something like this:

- "I have a ton of initiative. I'm not afraid to get my hands dirty or take a risk. At my last job, I felt a lot of non-engineers would benefit from understanding more about how the product is built. I started a weekly lunch program where engineers would give technical talks to non-technical people about the product, recent developments in technology, and other related topics. Almost everyone in the company came to at least one talk each month, and about 10 percent of the company attended every single talk."

- "I'm scrappy and find creative ways to get things done. For example, I wanted to teach a class in college, and undergraduates were not really allowed to do that. By partnering with a well-respected professor who would officially oversee the class but not get directly involved, I was able to teach something entirely on my own."

- "I think one of my biggest strengths is, interestingly, a lack of stubbornness. I can be passionate about my ideas, but I'm also not afraid to admit I'm wrong. This has helped me be flexible when aspects of the product or marketing changed quickly, and it's also helped build my coworkers' trust in me."

Do not list "intelligence" as a strength. Yes, you might be very intelligent. Yes, it is relevant to the position. However, listing this just comes across as arrogant, and, frankly, the interviewer will make her own decision as to whether or not you're intelligent.

For weaknesses, you want to give a genuine weakness. The old trick of disguising a positive attribute as a weakness ("I work too hard") does not work—and probably never has. Your interviewer wants to see that you can admit to your faults.

That said, it is a good idea to accentuate—or at least mention—the positive side of the weakness. What have you learned from it? Are there any benefits to this? How do you compensate for this weakness?

- "I'm often not very detail oriented. While this has some benefits, particularly in an entrepreneurial environment by helping me move faster, it also means I've made some embarrassing mistakes in the past. I've developed a good sense though of the types of situations where I'm likely to make mistakes, and I double or triple check my work in those cases."

- "I can sometimes come across as too negative or critical, even when I don't mean to be. I know I've hurt some people's feelings in the past, especially earlier in my career. I've worked on my communication style a bunch to mitigate this. I precede any negative feedback with positive comments, and I ask questions instead of directly asserting the criticism."

- "I can sometimes get distracted at work, particularly when I have a lot of different tasks on my plate. I try to do everything at once, get overwhelmed, and don't make solid progress on any of them. I've found a way to deal with this when I have multiple competing responsibilities by developing a clear to-do list. This imposes more structure on my workday and prevents the feeling of being overwhelmed."

If it makes you feel more comfortable or helps you accentuate the positive, you can phrase your weakness as "One of the things I'm working on is …." This can show you are working on your weaknesses and shift it around in your head to a topic you don't mind talking about.

Generally, try to avoid weaknesses that are too tactical or temporary. An answer like, "I don't have much experience in people management." doesn't help an interviewer get to know you and doesn't show humbleness. In fact, it just looks as if you're afraid to admit to genuine weaknesses—or don't think you have any personality faults.

Of course, candidates can go too far to the other side and offer up a deeply detrimental weakness, such as lack of honesty or a poor work ethic. This is very rare to see, though. When candidates flunk the weaknesses question, it's almost always because they came across as trying to hide genuine faults.

Below, we have provided a list of sample strengths and weaknesses. This is only to get you started; your strengths and weaknesses might not be on this list. You should pick ones that ring true for you.

Sample Strengths

analytical	scrappy	creative
energetic	organized	decisive
thinking outside the box	risk taking	calm under pressure
thorough	see things through	understand people's feelings
flexible	initiative	detail-oriented
good planner	quantitative	multitasking
leadership	good at taking feedback	persistent
persuasive	data-driven	independent
self-critical	good mentor; caring	not afraid of challenges
prioritization	enjoy learning new skills	add humor and fun to a team

Sample Weaknesses

not detail-oriented	overly confident	lack of confidence
too negative	makes too many assumptions	unrealistic
unassertive	impatient	indecisive
stubborn	intimidating to others	procrastinator
take feedback personally	difficulty admitting failure	hesitant asking for help
too direct / blunt	overanalyzing	argumentative
easily distracted	can be very vague	bad at multi-tasking
micromanages people	short attention span	shy

Behavioral Questions
Chapter 12

Behavioral questions come in many shapes and sizes. They could ask how you would respond to a hypothetical situation or how you actually handled a particular type of situation. They also may ask you to elaborate on a particular section of your resume. Ultimately though, they're all looking at the same two factors: your content and your communication.

Unfortunately, many candidates forget about one or both of these factors. They just try to come up with any answer that matches the question, failing to focus on what their answer says about them or how they deliver the answer.

Your goal is to master both of these aspects.

Why These Questions Are Asked

As we said, behavioral questions are about your content and your communication. Within these categories though, there are multiple factors.

Content

In assessing the content of your response, an interviewer might be trying to drill into the following areas.

Have you *really* done what you say you've done?

One important—and often overlooked—reason that interviewers ask behavioral questions is to validate that your resume really matches up with your experience. It's not that the interviewer necessarily thinks you're lying (although some candidates do); it's that it's often difficult to assess what someone did over weeks or months of work from a mere fifteen-word bullet.

Imagine a resume bullet such as, "Implemented new process that led to 15% increase in user retention rates."

This seems pretty amazing, but is it really? The new "process" might have been a matter of just including a renewal link in an email. It was still a very good change, of course, but a change like that probably doesn't indicate a ton about your future success.

On the other hand, the change might be a lot more impressive than your bullet made it sound. One or two lines doesn't offer you the chance to really explain why something was hard. Maybe implementing a new process required you to build a makeshift team, gather data, and perform complex analysis.

The fact is n interviewer can't bet their interpretation of a line on your resume matches what you actually did. Your accomplishment might have been a lot harder or a lot easier than you made it sound. Behavioral questions allow an interviewer to "dive deep" into your prior experiences.

How have you made an impact?

The scale of the situations you discuss also indicates something about how big your accomplishments have been and what you consider to be an accomplishment.

For example, suppose you're asked a question such as, "Tell me about a challenging problem you've faced."

That's actually a very open-ended question. Your challenges could be teamwork issues, or they could be analytical or quantitative problems. They could even be personal issues.

The scale and the type of the problems you select tells your interviewer a lot about the complexity of your professional history. This is especially true for questions about successes, failures, and challenges. If you haven't had professional wins *and* losses, then many would argue that you haven't really done much at all.

Do you have the relevant skills and attributes for the job?

In addition to the above elements, behavioral questions are typically used to evaluate your skills and attributes. An interviewer might have a very particular skill or attribute in mind, such as leadership skills, or they might be asking a broader question to gain a sense for how you handle things.

A question such as, "Tell me about a time when you've had to convince a manager of something," is looking at how and why you influence people when you lack power.

- Do you just go with whatever a manager says? If not, when do you speak up and why? Is it about your personal goals or those of a team? Is it about the short-term or the long-term goals?

- How do you influence someone when you don't have the power to just make something happen? Do you gather data to support your conclusion? Do you rally support from those around you? Do you bring up risks or opportunities? What tools did you use to influence your manager, and which of those were successful? How did you adapt to your manager's style?

Other questions might be broader. A question such as, "Tell me about a mistake you made," is likely looking at a bunch of factors. Such a question gives an interviewer a sense of how you handle situations, if you've faced real challenges, how you recover from mistakes, and what you see as a mistake. It *also* shows an interviewer (hopefully!) that you can be humble and admit when you screw up.

Communication

There's good news about behavioral questions: even if you can't nail the content part, you can at least nail the communication part. The trick is to apply some specific structures.

For people who struggle with communication issues, including non-native English speakers and those prone to rambling, this advice is especially important.

Nugget First

The "nugget first" structure is a simple one. It means to start off your response with the "nugget"—or thesis—of what your story will be about.

For example, when your interviewer asks you about a challenge you faced at work, you might open your response with, "Sure, let me tell you about the time had an underperforming teammate."

Such as opener helps your interviewer focus on what you're about to say, allowing them to slot all the bits of information you're about to give them into that context. The remainder of your story feels more structured as a result. When you mention additional details, such as being behind on a deadline, your interviewer knows that the point of the story is the coworker, not the deadline.

It also helps to focus *you* on what you're about to say. You'll be less likely to provide extraneous details because you know your answer is about the underperforming coworker.

Situation, Action, Result (S.A.R.)

The Situation, Action, Result structure can be used on its own or in conjunction with the Nugget First approach.

The S.A.R. (also often called S.T.A.R.—Situation, Task, Action, Result) approach means to describe the following components:

- **Situation:** Your goal here is to provide *sufficient* background information to understand what you did and why it mattered. Be careful about overwhelming your interviewer with information that doesn't truly inform them. For example, if you're describing a project where your team was behind, it may not be necessary to describe the details of the project. Simply saying "We were working on a project for a key client" might be enough.

- **Action:** You then describe the actions you took. Note that the story should focus primarily on what *you* did, not what your team did. Your interviewer, after all, is hiring you.

- **Result:** Finally, you explain the results of your actions. How did you help your team or your company? How did people respond? When possible, quantify your impact. Tell your interviewer your actions led to a 10% increase in user retention, not just "greater" user retention.

Suppose you're asked the following question: "Tell me about a time when you had to influence a team." You might have a story like this:

- **Nugget:** Sure, let me tell you about the time I needed to cut the scope of a project.

- **Situation:** To meet an important release deadline, we needed to cut a key feature. Unfortunately, this was a feature the developers were particularly excited about. If I just came out and told them, "Sorry, but no," they'd be demoralized and feel that their opinion wasn't valued.

- **Action:** I identified two developers as being particularly strong influencers, one because she was very senior and the other because he was very well liked. I focused on getting them on board with the decision first.

  I met with the senior dev first to discuss the impending deadline and what it meant for the product's architecture. We discussed options and consequences, and ultimately she arrived at the same conclusion: we needed to cut this feature.

  With the other developer, I told him about the discussion with the senior dev, appealing to this dev on more of an emotional level. I knew he valued his social status on the team, so I told him how much his support on this would mean to me. He shared his concerns about the team's reaction and assured me that he would support me.

  I asked both developers to help me present the decision to the team.

- **Result:** The team was ultimately disappointed in the conclusion but sympathetic. Hearing two highly valued team members back the decision, they trusted that it was the right one. I pledged to them that, if we could push forward to get this release done ASAP, we'd start working on this feature right after that release.

  We ended up making the deadline with a day to spare and then getting that other feature out a few weeks later.

By structuring your response this way, you allow the interviewer to clearly understand what was going on, what steps you took, and what the result was. Note that, in this case, there was no need to discuss what the feature was or why the developers were excited about it. It's not relevant to this story.

Speak In Bullets

Sometimes people joke that PMs speak in bullets, as though they're always doing a PowerPoint presentation. This isn't a bad thing; in fact, speaking in bullets can be a very good thing.

When you're speaking to an interviewer, or anyone, about an experience you had, they have far less context than you. It's easy for them to get lost in a myriad of details and forget who is who in your story. Being very clear about the components of your goals, tasks, or results will help your interviewer.

For example:

- "I did three things. First, I talked with … Second, I … And third, I …"

- "We had two issues with this plan. Number one, we … And number two, we …"

If you can picture what you say in a bullet list, and deliver it that way, your interviewer will have a much easier time dissecting the information.

Preparation

To deliver strong content, you need to be well prepared. In fact, in many ways preparing for behavioral questions is probably the best "bang for your buck." You know you'll be asked behavioral questions and, with proper preparation, you really can master these questions.

Step 1: Create a Preparation Grid

A preparation grid helps you map your experience to the common behavioral questions. It might look something like this:

	Job 1	Job 2	Extracurricular
Leadership / Influence			
Teamwork			
Successes			
Challenges			
Mistakes / Failures			

In the columns, you list each major "chunk" of your resume: each role, each project, and each extracurricular (volunteer, etc.) activity. In the rows, you list the major categories for behavioral questions: leadership / influence, teamwork, successes, challenges, and mistakes / failures. Then, fill in each cell with one to three stories.

Creating such a grid will make it easy for you to come up with an answer to a very specific question. When your interviewer asks you a question like, "Tell me about a time when you dealt with a hostile coworker." or, "How did you show leadership at your last company?" you'll have a story on hand to deliver.

Step 2: Master Five Key Stories

The cells above have lots of stories, but what you want to master is five key stories that best represent why you're a great PM candidate. These are the stories that you'll try to fit in whenever you have a chance. This is where you get to show off your wow-factor.

When you identify the five key stories from your past, ask yourself the following questions.

Is each story substantial?

Each of your stories should have a substantial situation, action, and result. Most often, stories flop in the "action" part.

Consider this story which I rehearsed with one candidate:

> My team had just launched a new feature for updating your profile from the mobile app. Everything seemed to go well at first. But then, several users began complaining they were getting spammed.
>
> I gathered the developers together and asked them to focus on figuring out if this was our fault. It turns out the process of updating their profiles had inadvertently changed their privacy settings.
>
> The developers were able to remedy the situation. They reverted back to old privacy settings and rolled out a fix.
>
> I then drafted an email to the users explaining the situation and our remedy. The users were happy with our resolution and the issue hasn't come up again.

When the candidate got to the end of this story, I nodded and said, "So you basically just wrote an email, right?"

"Well, yes," the candidate told me. "But it could have been a really big problem if we hadn't fixed it so quickly."

She's right; it *could* have been. It was the developers who tracked down the problem. All this candidate really did was write an email apologizing. That's not substantial.

Make sure each of your stories has a "meaty" situation, action, and result. Rehearse your story with someone and have them parrot back to you the gist of these three components.

Many stories that flop initially can be salvaged. For example, if this candidate had pulled together a new process to prevent these situations from occurring again, then that could well make a good story.

Is each story understandable?

Even a story with a great situation, action, and result will flop if it's too complicated. Your interviewer will wind up lost in the details about who did what and be unable to grasp the essential parts: what *you* did.

Of course, a good part of making stories understandable is how you deliver them. You need to understand what the gist is of each story so you can offer up just the relevant bits.

If you've stripped the story down to just the essentials and still find yourself explaining a lot of details, it might be time to kill the story. This is especially common with stories requiring technical, product, or mathematical knowledge.

What does each story say about you?

A story you give your interviewer shouldn't just answer the question; it should communicate something to your interviewer about you or how you work. When you review a potential story, ask yourself: what does this say about me?

For example, take the following story:

> *At my last company, I identified a way that we could, with a few tweaks, support a new market segment—startups. However, the CEO was reluctant to pursue this opportunity, explaining that he was worried about the additional development time and wasn't sure this new market would be lucrative. He just didn't want to take the risk for so little return.*
>
> *Rather than give up, I met with one of the key software architects to develop a plan to test out these new tweaks. We were able to reduce the development plan for a minimum viable product from six weeks to just two. We also worked out a way to minimize the effort required to support the market, reducing the marketing, testing, and support costs.*
>
> *I also researched the market more thoroughly. I explored what other solutions that market had and got a feel for how our product would compare to the other options. I then searched our customer base and found that about 5% of our customers were actually enterprise customers.*

CRACKING THE PM INTERVIEW | MCDOWELL & BAVARO

When I re-approached the CEO, I had a considerably less risky proposal. We could spend just two weeks of development costs and roll out this new version to just the existing enterprise customers. This would reduce our marketing costs to nearly zero, at least until we knew if we definitely wanted to go ahead and pursue it. And since we're not rolling out a new product to new users, it's easy for us to revert the changes.

The CEO agreed to this plan as a way to explore this opportunity with the new market segment. It ended up being very successful—in fact, it increased user spend by 20% for those users. This new market now accounts for about 15% of our total revenue.

This candidate has shown that she is:

- **Creative:** She identified a new market segment.

- **Ambitious:** She pursued an idea when others might have been discouraged

- **Understanding:** She understood what the CEO's concerns were and found ways to resolve them, rather than just arguing that he was wrong.

- **Data Driven / Analytical:** She gathered data to support her pitch.

- **Scrappy:** She found ways to reduce costs and analyze a market at minimal risk to the company.

You could even approach it from the other direction. What do you want to show about yourself, and how can you communicate that through examples?

- Are you creative?

- Are you well liked within your company?

- Are you good at building teams and energizing people?

- Are you understanding and empathetic?

- Are you a good leader?

- Are you analytical?

- Are you determined and ambitious?

- Are you a risk taker?

- Are you good at minimizing risk?

CrackingThePMInterview.com   175

- Are you technical?

- Are you good at influencing others?

Each of these areas could be part of a new story.

Is it really about you?

A good product manager is all about putting her team first.

This is a good thing to do in general, but many take it too far in interviews. Listen to the way you tell your story. If you find yourself constantly saying "we" and "us" instead of "I" and "me," you might want to tweak your wording—or find a new story entirely.

These stories should be able your successes and failures, not your team's. Your team is of course a key part of those (either enabling or feeling the impact of them), but it's still about how your actions contributed to those. For example, if your team pivoted, you should discuss what you did to lead your team through that pivot.

The actions should be your actions, but the results should be felt by the team or your customers.

Do you "get" other people?

Understanding other people is a fundamental part of teamwork, leadership, and persuasion, and therefore a fundamental part of a product management role. Your stories should show that you "get" other people.

Unfortunately, sometimes a candidate's stories reveal that she lacks insight into other people. This commonly happens when a candidate:

- **Bad mouths other people:** Yes, we all know some people can be frustrating to work with, but an interview is not the time to dive into this. If you rant about a coworker, your interviewer is likely to wonder if the issue was really that *you* are hard to work with. Many companies won't want to roll the dice there. Instead of complaining about the other people in the story, address their motivations. You might say, "The other PM was trying to optimize for immediate revenue" rather than "he was an idiot."

- **Shows helplessness:** Failure is okay; helplessness is not. If you weren't able to convince other people of something, explain why—without casting blame. For example, you might say, "the executives and I decided that while my proposal was exciting, changing course so late in the project introduced too much risk."

Empathy is an important part of any PM role, and these stories need to demonstrate that—or at least not demonstrate the opposite.

Do your stories have good coverage and good flexibility?

Behavioral questions tend to fall into one of five areas, so you'll want to make sure that you have good coverage of these topics across your five key stories. These common topics are:

- Leadership & Influence

- Challenges

- Mistakes / failures

- Successes

- Teamwork

You should have at least one story per topic.

Ideally, several of your stories will fall under two or more categories. This is valuable because you don't know what you'll be asked. What if you're asked multiple questions on challenges? Or what if a question doesn't cleanly fit into any topic? You might find that you've already used your "challenge" story for a question like "Tell me about a time when you led a team through something difficult"—only to get a question that explicitly calls for leadership.

By ensuring your stories have both good coverage and good flexibility, you'll be able to prevent getting caught off-guard with specific behavioral questions.

Step 3: Practice

You know that you'll be asked a variety of behavioral questions, so there's no excuse for not being well prepared for these questions. If possible, grab a friend to listen through your stories. Ask your friend to then repeat back what he understood the gist of the situation, action, and result to be. Did each part sound substantial to you and him? If he struggles to understand what your story is about, then your interviewer will likely as well. That means it's time to pick a new story or refine your current one.

Follow-Up Questions

Just because you've nailed the question doesn't mean your interview will let it go at that. Depending on the question you're asked, be prepared for follow-up questions such as:

- How did the team react?

- How did this affect the future of the team?

- What did you learn from this?

- What would you do differently if this situation came up again?

- Do you always handle situations like this? If not, what made you handle things this way this time?

Types of Behavioral Questions

We have listed some of the common behavioral questions below. Note that questions can take many forms and might be a great deal more specific than this. For example, rather than asking you for a challenge you've faced, you might be asked for a challenge you faced on a specific project.

Leadership & Influence

Leading and influencing others is perhaps the most essential responsibility of a PM, so it's no wonder these questions come up so often. A product manager, unlike a people manager, often needs to influence others without direct responsibility over them, so an interviewer will want to understand what tactics you use to build teams, persuade or influence others,

Some common tactics include:

- Gathering data to support your conclusion.

- Understanding and addressing people's underlying motivations or incentives.

- Developing support from key team members first and then leveraging that to get other people on your side.

- Showing your own vulnerability to encourage others to show theirs.

- Being a good role model or example.

- Gradually leading people to a conclusion by agreeing on a common framework first.

- Developing credibility and engendering trust.

If none of these jog your memory, think back to times when you've influenced other people, whether it was a coworker, a friend, or even a family member. What tactics worked for you? Or what tactics worked on the other side to influence you?

Sample Questions

1. Describe a decision you made that wasn't popular. How did you handle implementing it?

2. Describe a time when you had to motivate employees or coworkers.

3. Tell me about a time when you showed initiative.

4. Tell me about a time when you had to give a presentation to people who disagreed with you.

5. Tell me about a time when you had to make an unpopular decision.

6. Tell me about a time when you had to sell another person or team on your idea

7. Tell me about a time when you've built a team.

Challenges

Questions about challenges are less about the challenge (although it should be something meaningful) and more about understanding your reaction to challenges. After all, you will be confronted with challenges in your job. They want to understand how you solve problems.

To brainstorm challenges or problems you have faced, consider if you've encountered any of the following issues at work:

- Ethical dilemmas.

- Conflicting incentives.

- Insufficient resources (time, money, expertise).

- Incomplete or inaccurate information.

- Low morale, interpersonal issues, or other emotional problems (with teammates or the team as a whole).

- Cultural or workstyle conflicts.

- Changing demands.

- Inability to accomplish a task or meet expectations.

Since there are a wide variety of problems you could have faced, there are a wide variety of ways to solve these problems. Some common tactics include:

- Gathering data to decide what to do.

- Leveraging the support and expertise of people around you.

- Discussing and setting team priorities.

- Understanding the emotions of those around you.

- Thinking about what the "right" thing to do is (based on ethics, what's best for the customer, etc).

- Breaking down the situation, focusing on what you know, and understanding more about what you know.

- Mitigating risk.

- Being honest and straightforward.

- Solving the problem creatively or thinking outside the box.

- Compromising.

- Balancing short-term and long-term tradeoffs.

- Managing the expectations of coworkers and customers.

Some of these challenges could be from an activity outside of work, if you've done something substantial there. However, you should be prepared with stories from your formal work experience as well.

Sample Questions

1. Tell me about a time when you faced a challenge and overcame it.

2. Tell me about a time when you weren't able to reach a deadline.

---

3. Describe a major change that occurred in a job that you held. How did you adapt to this change?

4. Tell me about a time when you had to deal with changing priorities. How did you handle it?

5. Tell me about a time when you had to make a decision quickly or with insufficient data.

6. Tell me about a time when you used a lot of data in a short period of time.

7. Tell me about a time when you handled a risky situation.

Mistakes & Failures

At least one of your interviewers is likely to ask about failures or mistakes. The goal of this question, as with other behavioral questions, is to sell yourself. However, what "selling yourself" means here is a bit different than in other questions.

Your interviewer will be looking for the following:

- **A Big Failure:** Your interviewer wants to see that you've truly failed before. Right or wrong, many people believe that if you haven't failed then you haven't really tried. No one is perfect all the time. Thus, counterintuitively, you need to have a substantial failure under your belt.

- **Humbleness:** Your interviewer is also looking to see that you can *admit* failure. They know that you've failed at some point in your life. But can you admit it, even in a sensitive moment (such as an interview)? Or will you try to sugarcoat it? You want to come across as genuine and sincere.

- **Handling:** Your interviewer wants to see how you handled the situation. Did you correct the mistake for that incident, or do something to prevent future incidents? How did you relay the information to your manager and/or team? What did you learn from it?

One way to come up with good stories for this is to think about the things you've learned in your career; often there was a mistake that prompted the learning. For example, maybe you built exactly what the customer asked for, but they didn't use it. This taught you that you need to dig into the deeper motivations of the customer. Stories like this can be great because they naturally lead into the positive ending: you learned something important that makes you a better PM now.

A small number of candidates go too far, and their failure amounts to a "red flag." If your answer involves doing something grossly wrong and/or truly hurt someone, you might want to pick something else. There's a "sweet spot" of the size of your mistake. Most candidates are on the "too weak" side, but a small number are on the "whoa! too much!" side.

In particular, avoid answers that cut into your honesty or integrity. It's okay to have made a mistake analyzing data or trying to take on too much. But if you lied or cheated, that's a much harder thing to recover from.

Sample Questions

1. Tell me about a time when you made a mistake.

2. Tell me about a time when you failed.

3. Tell me about a time when you improperly analyzed a situation.

4. Tell me about a time when you were disappointed with yourself.

5. Tell me about a time when you were unable to juggle all your responsibilities.

Successes

Success questions are your time to shine. Your answer to this question might well overlap with your answers to some of the other questions, and that's okay. Your success might be a big challenge you conquered, a time when you led a team to success, or even a situation where you overcame a potential mistake or failure.

For these questions, think about the things that you are most proud of. Why are you proud of this action? Is it because it was hard? Because it had a huge impact on your company? Because it was outside of your comfort zone? Any of these aspects can make a fantastic answer. Pick something that is meaningful to you.

Make sure you can explain to the interviewer why you see this as a success. If it's because it had a big impact on your company, you should be able to quantify the impact. If it was outside your comfort zone, this should have changed you in some small way or helped you discover something about yourself. If it's about solving a hard problem, you should have learned something.

Sample Questions

1. Tell me about something you're proud of accomplishing.

2. Tell me about a time when you reached a goal that was important.

3. Tell me a specific insight you gained from something outside of work.

4. Tell me about a time when you went above and beyond the call of duty.

5. Describe a time when you resolved a situation before it became serious.

6. Tell me about a time when you had to show innovation.

7. Tell me about a time when you solved a problem in a creative way.

Teamwork

Teamwork questions are used to assess your interpersonal skills, particularly in times when you are working with your immediate peers. Look for times in your work experience where interpersonal communication or differences in work style affected your team dynamic.

How might you have solved these issues? Potential ways include:

- Striking compromises across people.

- Finding ways of making teammates feel valued.

- Being able to agree to things that are suboptimal for you in the interest of the greater team good.

- Understanding people's underlying motivations and incentives.

- Motivating teams and boosting morale.

- Relinquishing your ego and encouraging others to do the same.

- Setting common goals, metrics, and procedures.

- Balancing autonomy with team cohesion.

- Building the confidence of those around you.

- Increasing individual accountability.

- Setting a good example.

- Taking personal responsibility.

- Showing compassion and empathy for coworkers.

- Identifying and dividing responsibilities.

- Sharing knowledge and responsibilities.

- Mitigating the damage from a negative teammate or situation.

- Building trust across the team.

There is no "right" way to foster positive team dynamics. Part of delivering an excellent answer to this question is understanding that good teamwork depends on the specific situation.

Sample Questions

1. Tell me about a time when you had to work across teams to accomplish something.

2. Tell me about a time when you had a disagreement at work.

3. Tell me about a time when you mentored or aided a coworker.

4. Tell me about a time when you had to do something you didn't want to do.

5. Tell me about a time when you had to compromise.

6. Tell me about a time when you had to resolve a conflict.

7. Tell me about a time when you had a challenging interaction with a coworker.

Estimation Questions
Chapter 13

I promise you: no one actually cares if you know how many pizzas are sold every year in Manhattan. Even if you knew the "right" answer, it wouldn't help you. In fact, it could distract you. This is one case where the correct answer isn't necessarily the best one.

Estimation questions are entirely about the process you take to solve them. It's the journey, not the destination, so to speak. Interviewers will use these questions to evaluate your problem-solving skills as well as your quantitative skills.

What's the relevance to PMing, you might ask? Quite a bit, actually.

Other than the obvious (an ideal PM is good at problem solving and good at math), being able to estimate things is actually a valuable skill. After all, when you need to figure out what you might be able to expect for the revenue from a given feature, estimations will come in handy.

In life, as in the interview, you'll be shooting for a number in the right ballpark. Precision is not expected.

Approach

Since estimation questions are fundamentally problem-solving questions, it should be no surprise that the key to these questions is the approach. You can ace these questions with a bit of structure (and some tips and tricks).

Step 1: Clarify the Question

You can't answer a question that you don't understand. That's why it's important to make sure you heard the question correctly and truly understand what's being asked.

For example, if you're asked how much money Gmail makes in ads every year, you'll first want to repeat the question back to your interviewer. This is especially important if you're a non-native English speaker or if you think you might have misheard the question.

Next, you'll want to ask about anything that's ambiguous. For example, in the above question, there's actually a lot that's ambiguous:

- Does "money" mean profit or revenue?

- If it's profit, what are we including as costs? Do salaries count as a cost? How do we take into account the salary for a person who is only working on Gmail some of the time?

- Does Gmail include just Gmail.com? What about when companies use hosted Gmail for their domain?

- Do you mean the past year? Or averaging over all years since Gmail was launched?

- Is this for the US only, or is it worldwide?

The answer to each of these questions would substantially affect your approach and result.

Step 2: Catalog What You Know (or Wish You Knew)

Once you understand what the question is, you'll want to get a feel for what knowledge you have and what you need to compute. In some cases, you can ask your interviewer for key facts.

Because there are multiple approaches that can be effective, what you know can guide your approach.

For example, if you're computing the revenue from Gmail annually in the US, you might know (or be able to ballpark) some of these facts:

- The population of the US.

- The percent of people in the US who own a computer.

- The unemployment rate in the US.

- Google's annual revenue.

- The cost-per-click of a typical advertisement.

- The click-through rate of an advertisement.

- How well a Gmail or other embedded ad does as compared with a search advertisement.

- The number of ads shown on Gmail.

You might be able to ask your interviewer for some of these facts, but be prepared to compute some of these too. If it's something that many candidates would know and you don't (like the US population), then it's probably fair game to ask.

When in doubt, leave your question open-ended: "Could you tell me the click-through rate of an advertisement, or would you prefer that I compute it?" If your interviewer pushes back by saying something like, "What do you think it is?" that's a good sign that you might be relying on questions too much.

Steps 2 and 3 are done somewhat in parallel. You can come back to your Fact List as you're developing your equation.

Step 3: Make an Equation

This is possibly the most important part of the whole process. Here is where things come together—or fall apart.

You need to form an equation to solve the problem. Doing so will help you make progress and show your interviewer that you can tackle tough problems head on.

There are multiple approaches for any problem, but some are better than others. However, there's often no "best" approach, as it depends on your knowledge and background.

If you are calculating the annual revenue for Gmail in the US, this approach might work well:

```
[# Gmail users in the US] x [annual clicks per user] x [average
revenue per click]
```

If you think about the above values mathematically, the components will multiply together to give you the revenue.

This isn't the only approach though. You could also try something like:

```
[Google US revenue] x [% revenue from ads] x [% ad revenue from
embedded (non-search) ads] x [% embedded ad revenue from Gmail]
```

Before trodding down the path of one equation, it might be useful for you to brainstorm multiple equations and to have a plan of attack for each component. You don't want to waste a bunch of time computing one part and then realize you aren't able to compute the next one.

Step 4: Think About Edge Cases and Alternate Sources

Pause for a moment and think about the potential problems in your approach. This is a great opportunity to show your interviewer that you're detail oriented and not afraid to challenge yourself. No one wants to hire someone who will just brush problems under the rug.

For example, if you're computing the number of pizza places in the US, have you considered college towns? Or, if you're computing how much it would cost to wash all the windows in your city, have you taken into account broken windows (which you presumably won't wash)? Or even car and bus windows?

Is there some source that you haven't considered? For example, if you were computing the number of guns sold every year, have you included illegal sales? Sales to police as well as consumers? What about race tracks using blanks?

Think carefully about what situations might not fall into the framework you created in the previous step.

Some candidates worry about exposing faults to your interviewer. Don't! Your interviewer has likely asked this question dozens of times. He'll know where the flaws are whether or not you point them out.

Step 5: Break It Down

At this point, you have a strong framework solving the problem. You have an equation written which lead you to an answer. You've even analyzed potential edge cases and issues.

Now you want to find a way to solve each component of the equation. Let's take a problem like computing the annual Gmail revenue in the US and assume we're working off this equation:

```
[# Gmail users in the US] x [annual clicks per user] x [average
revenue per click]
```

How can we compute the number of Gmail users in the US? If the interviewer won't give us this information, we can compute it through an equation (yes, another one!) like this:

```
[# Gmail users in the US] =
      Population of the US
   x % of People with email
   x % of email users who use Gmail
```

Perhaps you recently heard something on the news about the number of Hotmail users. In that case, you could consider something like this:

```
[# Gmail users in the US] =
      # Hotmail users x Ratio of Gmail users to Hotmail users
```

Even an equation like this might work:

```
[# Gmail users in the US] =
      # Smartphone users x [
      % Android marketshare * % Android users w Gmail +
      % iPhone marketshare * % iPhone users w Gmail +
      % Blackberry marketshare * % Blackberry users w Gmail +
      % Windows Phone marketshare * % Windows phone users w Gmail
   ]
```

The last equation might seem a bit crazy, but there's some logic behind it: Gmail usage might be correlated with income, just as smartphone usage is.

For each part of the main equation, construct a sub-equation. Keep each part of the original equation separate; don't try to merge them into one mammoth equation. You might even draw a line down the page (or whiteboard) to keep these computations entirely separate.

Step 6: Review & State Your Assumptions

We now have a series of equations and all that's left to do is guess some numbers. You'll need to rely on your intuition and prior life experience. You might have an idea of how often you see Gmail addresses versus other addresses.

Be aware of your biases. Yes, everyone that *you* know has a smartphone, but is this really representative of the entire US? Probably not.

Pick nice, round numbers, and state your assumptions clearly to your interviewer. Keep the list of assumptions you made in Step 2 updated, or find some way to clearly identify where you've made an assumption.

Be sure to tell your interviewer *why* you're making the assumptions you are. Your reasoning is more important than the precise number.

Step 7: Do the Math

All that is left is to actually multiply these values out. Remember that we're shooting for the right ballpark, not absolute precision. Keep your numbers nice and round to make them easier to deal with.

Step 8: Sanity Check

Whew! You finally have an answer. Give that answer to your interviewer, but make sure you double check your work right after that. Make sure the value passes some quick sanity checks.

For example, suppose you compute that Gmail's US revenue is 5 billion dollars. Does that sound right to you? This would indicate that Gmail is making $16 *per person in the US*. This would be shockingly high given that most of the US doesn't use Gmail.

Go back through and check your work. If there's an issue, it's probably in one of the following areas:

• Original equation.

• Assumptions made.

• Arithmetic.

It may help to do quick sanity checks on each part of the problem as well. If you found that Gmail had 150 million users in the US, this is a problem spot. It's reasonable to believe that 50% of the US does not use Gmail.

Numbers Cheat Sheet

While these questions are not a test of facts, there are some very useful facts to remember.

If there's something else you need to know, you can ask your interviewer. In many cases though, you'd be expected to deduce the value.

Approximate Value	Data
300 million	US Population
3	Average People per Household (US)
100 million	# Households in the US
80 years	Life Expectancy (US)
65 - 70 years	Life Expectancy (World)
7 billion	World Population
700 million	European Population
4 billion	Asia Population
9000	Hours in a Year
500,000	Minutes in a Year
---	Company Revenue
---	Company Profit
---	# Users

Note how all of these values are nice, round numbers. This is to make them easier to work with.

Prior to your interview, try to look up the company's revenue, profit, and user count. Of course, for some startups, some (or all) of these values could be zero.

Tips and Tricks

Estimation questions are heavy on math. Whether you're good at mental math or not, these tips and tricks can make your life a bit easier.

Tip: Round Numbers

There are times when being a perfectionist and being detail oriented is warranted, but this is not such a time. After all, you're going to be taking so many wild guesses and "guesstimates" that a bit of hand waving in your math really won't make a difference. You're just trying to get your answer in the right ballpark. Absolute precision will not do you any favors.

Example

- The US population is 314 million (in 2012), but you should use 300 million

- There are 8760 hours in a year (8784 in a leap year), but using 9000 or 10,000 is probably close enough.

Trick: Rule of 72

Here's a fun and useful tip: if you need to calculate how long until something doubles, divide 72 by the percent increase.

That is, an investment, population, salary or other value increasing at x% per year will double after approximately 72/x years.

This rule works fairly well (being within 5 or 10% of the true answer) for smaller values of x. However, even up through a 100% annual increase (which is doubling within a year), the result is still within 30% of the actual answer.

- x < 20%: result is within 5% of actual answer.

- x < 65%: result is within 20% of actual answer.

- x < 100%: result is within 30% of actual answer.

- x > 100%: result gets increasingly less accurate. However, this means the result doubling in less than a year.

Of course, dividing 72 by something without a calculator is generally tricky. Using 70 or 75 will make your life easier and will generally be close enough.

Example

Salary Increase

- **Interviewer:** "A recent college graduate makes $65,000 per year. If she gets a raise of 9% per year, how long until her original salary has doubled?"

- **Candidate:** "We can apply the Rule of 72 here. 72 divided by 9 is 8. So it will take about 8 years."

- **Answer:** The precise answer would be that it takes 8.05 years for her salary to double. After 8 years, her salary would have increased by 99%. Not bad for a bit of rounding!

Population Growth

- **Interviewer:** "The 2012 census determined the US population to be about 300 million and increasing at a rate of 0.7% per year. If the rate holds steady, how long until the US population reaches 600 million?"

- **Candidate:** "We're basically looking for the length of time before the US population doubles, so we can apply the Rule of 72 by dividing 72 by 0.7. What is that? Well, it's certainly going to be between 72 and 2 x 72 (since 0.7 is between 1 and 1/2). We can guess about 100 years then, which would tell us that the population will double around year 2112."

- **Answer:** Even with all this "loose" math, the candidate came fairly close. If he had divided 72 by 0.7, he would have gotten 103 years. The precise answer is 99.3 years. The candidate's answer was reasonably close to both values, but even more importantly, he took a logical, quantitative approach. That's what these problems are all about.

Trick: Orders of Magnitude

When multiplying two large numbers, it's easy to make a mistake. This isn't a big deal if you're off by something in the 1s digit, but if you add an extra zero or forget a zero, you can wind up off by 10x or more. That *is* a big deal.

It's therefore useful to validate that your numbers are at least in the same order of magnitude as the expected result. One way to do this is the following: if you multiplied a and b together to get n, then the sum of the number of digits in the terms should be within one of the number of terms in the result.

Or, more precisely:

    digits(a) + digits(b) = digits(a * b)

Or:

    digits(a) + digits(b) = digits(a * b) + 1

For example, 823 * 1032 will have either 6 or 7 digits. (In actuality, it has 6 digits.) If you wind up with a number like 84936, you'll know you've done something wrong.

Tip: Be Confident

How many times have you heard someone say, "Oh, I suck at math" or "Numbers are just not my thing"? Whether it's true or not, an interview is not the time for such proclamations.

Companies want to hire PMs who are confident—or who can at least pretend to be. Don't let it show if you're intimidated, and *definitely* don't tell your interviewer that you're bad at math. Self-deprecating comments won't do you any favors.

Tip: Label Your Units

Many people think of labeling units of a value (e.g., "4 meters") as one of those pesky things that school teachers insisted on, but this habit will be invaluable here.

A common mistake that candidates make is getting confused in their units. They write down "4" to indicate "4 meters," but when it comes time to do some arithmetic, forget that it's not in kilometers. And, unfortunately, this can be a very difficult "bug" to find.

If you label all values with their units, you'll be much better off in the long run.

Tip: Consider Your Sources

Imagine you're asked to compute how many bags of potato chips are sold every year. You will want to make sure you understand the different *sources* of sales. Potato chips are sold in several places:

- In stores, direct to consumers.

- In vending machines, direct to consumers.

- From distributors, to schools, hospitals, movie theaters, etc.

- And many other places!

Multiple "sources" are common in market-sizing or revenue questions, but they can also come up in other ways.

For example, suppose you were asked how many elevators were needed for a one-block long, one-block wide, 20-story building. You'll probably chug along doing some work with the number of people in the building and how often they need to use the elevator. But have you thought about the freight elevator? That's a "source" too!

It's okay in many cases to decide that your additional sources can be ignored, but you still want to call them out to the interviewer. This shows an attention to detail.

Tip: Keep Discrete Steps Discrete

The more organized you can be, the better. If your estimation question has several independent steps, you should try to compute them separately. If (or when!) you make a mistake, you'll be able to come back to it, narrow in on the exact mistakes, and correct it with minimal hassle.

Example

Suppose that you are computing the amount of revenue Gmail earns in ads every year. Your approach might look like the following:

```
$ = [# of Gmail users] x [# clicks / user] x [$ / click]
```

When you work through this problem, you should compute each of these entirely separately. You might have a chart that looks like this:

$ / click	# clicks / user	# of Gmail users
...	. . .	...

You'll merge (or multiply) these values only in the very last step in the question.

Tip: Record Intermediate Steps

While some estimation questions are very short, most are fairly lengthy and require numerous calculations. It's important that you write down what you're doing as you're doing it. You might need to come back to it later to correct your work or even to reuse a previously computed number.

Record each step in an organized, easy-to-read fashion such that you can easily come back to any step. Think of this as "showing your work" to your math teacher. If your approach is so organized that another person could hypothetically follow it, it will be much easier for you to analyze where you might have gone wrong.

Tip: Record Your Assumptions

It's not uncommon for your final answer to be wildly off. When this happens, there are two main reasons: either you made a math mistake (in which case being organized will help you locate it) or one of your assumptions was wrong.

Therefore, the easier it is to identify where you made an assumption, the easier it will be to discover potential issues.

Ideally, you would keep a list on the side of the page with all the assumptions you make. But, if that doesn't work for you, at least circle each assumption as you make it. That will go a long way in letting you skim all your assumptions for issues.

Example Interview

Before practicing some questions on your own, it might be useful to understand how an interview like this might be conducted. We have provided a sample dialog below. We have added section headlines for ease in reading and to make it clearer where the candidate is applying certain steps.

Interviewer: I'd like to start off with some estimation questions. How much money does the shampoo industry earn each year in the US?

Clarifying the Question

- **Candidate:** Hmm. You'd like to know the money earned by the shampoo industry annually, correct? Are you looking for profit or revenue?

- **Interviewer:** Let's do revenue.

- **Candidate:** Okay, great. And are we looking at just shampoo sales? Or are we considering conditioner and related products too?

- **Interviewer:** Just the shampoo is fine.

- **Candidate:** Alright. And when we say revenue, I'm assuming that we want to know how much the shampoo companies earn. That is, we're not including the profits earned by resellers. Correct?

- **Interviewer:** Exactly.

- **Candidate:** Okay, great. I think I have what I need to proceed then. So just to reiterate, we're looking for the total revenue pulled in annually from shampoo sales alone. May I have a few moments to jot down some thoughts?

- **Interviewer:** Sure, that's fine.

Catalog What You Know (or Wish You Knew)

Candidate records the following data:

- Population of the US is 300 million.

- Life expectancy is 80 years.

- How much shampoo does each person use per wash?

- What is a store's mark-up per item?

Make an Equation

Candidate comes up with one possible equation:

```
[# people in the US] x [# shampoo bottles used per year per
person] x [revenue / bottle]
```

- **Candidate:** Well, one way of doing this is to compute the number of bottles a given person goes through per year and then multiply that by the revenue per shampoo bottle. That will give us the annual revenue per person annually. We can then multiply that by the US population.

- **Interviewer:** Okay. Great.

- **Candidate:** I think that's a fairly good approach, but we'll want to break that down a bit more by market segment. Shampoo usage varies by gender as well as by age. Women, for example, spend more on shampoo than men since their hair is longer and since they tend to buy more expensive shampoo. So let's tweak our approach a bit by calculating these separately for men and women.

For women, we have the following table:

# women in the US	# shampoo bottles per woman	revenue / bottle
50% * US population	# showers per year / # showers per shampoo bottle	price per (female) bottle * (1 - store mark-up)

For men, it'll probably be easiest to just take what we've calculated for women and reduce it by a constant factor.

# men in the US	# shampoo bottles per man	revenue / bottle
50% * US population	# shampoo bottles per woman * (ratio of male shampoo bottles to female shampoo bottles)	price per (male) bottle * (1 - store mark-up)

We could further break this down into "luxury" shampoo users vs. normal shampoo users, but I think this is close enough.

Review & State Your Assumptions

- **Candidate:** I probably use about a teaspoon of shampoo per shower, and I think there's probably about 100 of those in a standard 12 ounce bottle.

  That should handle the middle part of the women's equation.

  For the right side of the equation, we first need to know the price per bottle. I think shampoo usually runs about $5 / bottle for women, and probably pretty similar to that for men. I believe stores mark up products by about 50%, so we'll work with that.

Candidate adds the above bolded parts to the assumptions list.

- shampoo per shower (woman): 1 tsp.

- ratio of male shampoo usage to female: 50%.

- size of shampoo bottle: 12 ounces.

- tsps per 12 ounces: 100.

- price per (female or male) bottle: $5.

- store mark-up: 50%.

Do the Math

- **Candidate:** From here, it's basically just a matter of plugging in these values and doing the math.

  For women:

# women in the US	# shampoo bottles per woman	revenue / bottle
50% * US population	# showers per year / # showers per shampoo bottle	price per (female) bottle * (1 - store mark-up)
50% * 300 million people	(365 showers per year) / (100 showers per bottle)	$5 per bottle * 50%
150 million	3.5 bottles / year per person	$2.5 per bottle

This is about $9 per woman, and therefore about $1350 million—or $1.35 billion—on shampoo.

Men have shorter hair, so we should expect that they use less shampoo. Let's say men use about half as much shampoo as women. This means that men contribute about $650 million to shampoo revenue.

Together, this is about $2 billion on shampoo.

That doesn't seem wildly off, but I'd like to double check some things.

- **Interviewer:** Okay, go ahead.

Sanity Check

- **Candidate:** Well, here's one potential issue. I calculated that women use 3.5 bottles of shampoo per year, where one bottle is about 12 ounces. This means that women are only using 3.5 ounces per month. This seems a bit low. When I travel with a travel-sized shampoo container, which is just about 3.5 ounces, this only lasts me about two weeks. I think my number for shampoo usage, for both men and women, is off by a factor of two. If I multiply this through, I'll end up doubling the revenue. The annual revenue should be closer to $4 billion.

- **Interviewer:** Interesting. Okay. Is there anything else?

- **Candidate:** Well, come of think of it, I think I made an inappropriate assumption before. I assumed that everyone's shampoo usage matched that of a typical adult (of their own gender), and I'm really not sure that's true. I haven't taken into account younger children or elderly / disabled people, or bald people. For that matter, I haven't taken into account that there's a bunch of people who just don't wash their hair every day. We need to adjust the usage down quite a bit for this.

Children below age 10 and the elderly / disabled account for about 20% of the US population. They will use far less shampoo than an adult (due to shorter hair or showering much less); let's just assume it's 0. This will account for 20% reduced shampoo usage.

Let's say that 20% of adult men are bald, so this accounts for 10% reduced shampoo usage.

Now, what about people who don't shower daily? Let's assume that 20% of people don't shower daily, and they instead shower every other day. This drops our shampoo usage by another 10%.

If we add these together, we find that we need to reduce our shampoo usage by 40%. So instead of $4 billion, we're back to a little over $2 billion in annual revenue.

- **Interviewer:** Excellent. Great work!

Sample Questions

Here are ten estimation questions which are similar to what you might see in an interview. We have provided solutions for these problems, but you should think of these solutions as being just one possible solution. There are often many correct solutions. Still, reading the solutions—after you've solved the problem yourself, of course—can be a useful way to explore alternative approaches or to remind you of details you might have missed.

Many of these questions are ambiguous, and that's okay! Part of your goal as a candidate is to resolve ambiguities prior to solving the question.

To practice these questions, make appropriate assumptions to resolve any ambiguity. In the solutions, we might make different assumptions. That's okay; our solutions are not *the* answer, but rather just one reasonable approach.

Where possible, we've obtained the "correct" answer from census or other data. We've provided this just for fun, but don't worry too much if your answer (or our answer) is far off. It is useful to understand why your answer was far off—was it a logical error or an unreasonable assumption?—but the actual accuracy doesn't matter.

The first three questions are tackled with a more detailed step-by-step approach so that you can get the hang of things.

Question 1: How much does the US spend on dog food each year?

We'll walk through this question step-by-step.

Step 1: Clarify the Question

What might be ambiguous in this question? Let's think.

- Are we including wet food and dry food? (Assume yes.)

- Are we referring to the end consumer here or stores? (Assume end consumer.)

There isn't a whole lot that's ambiguous here, so we can go ahead and proceed

---

with these assumptions.

Step 2: Catalog What You Know (or Wish You Knew)

We know or wish to know the following:

- The US population is about 300 million people.

- Average people per household is 3.

- How many dog owners are there?

- A large dog eats once or twice a day, finishing most of a typical sized dog bowl.

- A medium-sized bag of dog food is about 20 lbs.

- How much does a bag of dog food cost?

- Most dogs eat dry food.

- Of people who own dogs, how many dogs do they own?

Step 3: Make an Equation

An equation like this will probably work:

```
[# dogs in the US] x [amount of dog food eaten per year] x [cost
per unit]
```

We should pause here and think through this approach. Do we have a plan of attack for how we'll compute each of these components? Are there any other approaches that could work?

Step 4: Think About Edge Cases and Alternate Sources

Our approach has mainly centered on the number of pets. Are dogs used anywhere else? Sure!

- Police forces.

- Race tracks.

- Farms.

These probably don't make a big difference though since the number of dog owners is so large.

Step 5: Break It Down

Computing the number of bags of dog food consumed per dog and the price per bag will probably be pretty straightforward. Computing the number of dogs is a bit trickier.

We could just take a guess at the percent of American households with dogs, but it might be better to break down the number of households a bit better.

Here are a few ways we could break down the number of dogs in the US:

- Divide households by income brackets.

- Divide households by suburban, urban, and rural.

- Divide households by age: 18 - 30, 30 - 60, 60+. We haven't included children under 18 in there since they will be living with an adult.

- Divide households by apartment vs. house.

- Divide households by children: with children and without children.

Any of these (and many others) could work. You could even combine approaches—divide by age and then by income bracket—but that's probably making it a bit too complicated.

We'll take the last approach and segment the problem into households with children and households without:

# dogs in the US	amount of dog food eaten per year	cost per unit
[# households] x [% of households with dogs] x [dogs per dog-owning household]	[daily portion size] x [days per year]	cost per bag / size of bag

```
% households with dogs
    = [% households w kids] x [% kid households w dogs]
    + [% households without kids] x [% no-kid households w dogs]
```

Step 6: Review & State Your Assumptions

- Assume about 30% of households have children. I am guessing this because about 90% of people have kids at some point and kids take about 20 years to grow up. So this should mean that someone with kids at home has them for about one third of their adult life.

- Assume 30% of households with kids have dogs.

- Assume 10% of households without kids have dogs.

- Assume that 20% of households with dogs have 2 dogs. This means that, of households with dogs, there are 1.2 dogs per household.

- Assume that a negligible percent of families have more than 2 dogs.

- Assume the average dog eats about 1.5 cups of food per day. I've always owned larger dogs and they've eaten about 2 cups a day. So, small dogs—which probably eat proportionally to their size—should be eating around 1 cup per day.

- Assume a typical bag of dog food is about 20 pounds and costs about $15.

- I think 20 pounds of dog food is probably about 20 cups.

Some of these assumptions might be a bit off, but that's okay. Your interviewer isn't testing you on your knowledge of dog food.

Step 7: Do the Math

It's now just a matter of plugging in numbers. Remember to keep the units straight.

```
% households with dogs
    = [% households w kids] x [% kid households w dogs]
    + [% households without kids] x [% no-kid households w dogs]
    = 30% x 30% + 70% x 10%
    = 9/100 + 7/100
    = 16%
```

# dogs in the US	amount of dog food eaten per year	cost per unit
100 million households x 1.2 dogs / dog-household x 16%	1.5 cups / day per dog * 365 days / year	$20 per bag / 20 cups per bag
20 million dogs	~500 cups / year per dog	$1 / cup

If we multiply these out, we get $10 billion in dog food.

Step 8: Sanity Check

Does this sound right to you? Some ways to sanity check it are:

- Does $500 per year in food costs per dog make sense to you?

- Does 16% of households owning a dog make sense?

- If there are 20 million dogs per year, then there are about 15 people for every dog in the US. Does this sound right to you?

If there are major issues with this answer, where might they have come from? Here are a few ideas:

- We assumed / calculated that 30% of households have children. This probably isn't wildly off (it's not going to be 5% or 80%), but I could see the actual answer being anywhere from 20% to 60%.

- We assumed that 30% of households with kids have dogs. This was a pretty arbitrary guess, and therefore subject to issues.

- We assumed that 10% of households without kids have dogs. Again, this was a pretty arbitrary guess. It might be substantially off.

- We assumed that 20% of families with dogs have two dogs.

Brainstorming potential issues, and correcting your work if necessary, will be a valuable thing to do in an interview.

Actual Answer

There are 78.2 million dogs in the US[1]. We were off by a factor of about three, which isn't terrible. It's still the right general ballpark. Where did we go wrong though? The Humane Society provides some useful stats:

- 39% of households own at least one dog. (We estimated 16%.)

- 60% of dog owners own just one dog. (We estimated 80%.)

- 28% of dog owners own two dogs.

- 12% of dog owners own three or more dogs.

- On average, dog owners own 1.69 dogs. (We calculated 1.2 dogs.)

---

1   "Pets by the Numbers." Human Society. 27 September, 2013.

It looks like we underestimated the percent of households with dogs (16% overall vs. 39%), causing a difference of factor of two. We also slightly underestimated the number of families with multiple dogs. We calculated 1.2 dogs per dog owner, when there are actually 1.69 dogs.

Additionally, our food costs were off by a factor of 2. PetFinder estimated that dog food costs average about $250 per year[2]. We calculated $500 per year.

Americans should spend about $20 billion on dog food and we calculated $10 billion.

Question 2: How many tennis balls can fit in a two bedroom apartment?

This question is similar to a "classic" Google interview question: how many golf balls can fit in a 747 airplane?

Step 1: Clarify the Question

Like many questions, this one has some ambiguous or unstated components.

- Are we asking about a particular two-bedroom apartment (which could be of any size really), or a typical one? (Assume we're thinking about a typical apartment.)

- In order to understand what a "typical" two-bedroom apartment means, we need to know the location. Is this in the US? Any particular part of the US? (Assume a typical two-bedroom apartment in a downtown US location.)

- Are we asking how many tennis balls at most *could* fit in a typical apartment? Or are we asking about how many *would* fit we just threw them in at random? (Assume we're asking about the max that could fit.)

- Does the two-bedroom apartment have furniture in it? How much? What about people? (Assume a typical amount of furniture, but no people.)

- Can we put tennis balls inside places like cabinets? (Assume no. All the tennis balls will be "in plain sight.")

Step 2: Catalog What You Know (or Wish You Knew)

The things we know, or might wish to know, include the following:

2 How much does owning a pet cost in a year?" Kay, Liz F. 12 February 2012. Bankrate.com.

- The size of a typical two-bedroom apartment, in my experience, is about 800 sq. ft.

- A typical two-bedroom apartment has two bedrooms, a living room, a kitchen, and a bathroom.

- The furniture includes: two queen beds, two dressers, four side tables, a couch, a loveseat, a dining table with four chairs, one coffee table, and a TV.

- How much space do kitchen appliances take up? What about bathroom appliances?

Step 3: Make an Equation

Our basic equation will be fairly straightforward:

```
([volume of apartment] - [volume of furniture] - [volume of appli-
ances] - [volume of personal items]) / [volume of a ball]
```

Step 4: Think About Edge Cases and Sources

We've already gone through the different furniture and appliances in the apartment, so that takes care of a lot of potential issues. Another potential edge case could be a messy person who keeps their belongings in public view, rather than tucked away in cabinets and drawers.

However, since we're looking at typical usage, these edge cases shouldn't affect anything.

Step 5: Break It Down

We now need to break down each part of our equation.

Volume of Apartment

Since the square footage of an apartment is measured by the *interior* area, walls or number of rooms should not affect the volume.

```
volume of apartment = ceiling height x area of floor
```

Volume of Furniture

The furniture in a typical two-bedroom apartment will include:

- 2 queen beds (each of which has a mattress and a box spring)

- 4 night stands.

- 1 dining table.

- 4 chairs.

- 1 couch.

- 1 loveseat.

- 1 TV.

- 1 coffee table.

- 2 dressers.

We can probably ignore items that are primarily just a frame (such as a table), since they won't substantially impact the usable space. This reduces our list of furniture to just: 4 queen mattresses (the bed frame has negligible volume), 1 couch, 1 loveseat, and 2 dressers.

```
volume of furniture =
      4 * volume of queen mattress
   + 1 * volume of couch
   + 1 * volume of loveseat
   + 2 * volume of dresser
```

Volume of Appliances and Built-In Furnishings

A typical apartment will include:

- Bathtub.

- Sink / bathroom cabinet.

- Toilet.

- Kitchen cabinet units (including refridgerator).

We will assume that the volume of the bathtub and toilet is negligible. We do need to consider the sink and cabinets though.

```
volume of appliances =
     volume of sink / cabinet
   + volume of kitchen cabinets
```

Volume of Personal Items

A person's most significant personal items by volume are probably their clothing. Some will be put away in a dresser (in which case it doesn't affect the usable volume), and other stuff will be in the closet.

```
volume of personal items =
    % items in closet
    * volume of personal items
```

Volume of a Ball

Many people approach this by using the volume of a sphere. This is not quite correct. Unless we're grinding up the tennis balls into dust (and apparently filling the air inside the ball with something first), the volume of a sphere doesn't matter. The volume of a *cube* does. After all, the balls have gaps between them.

If the balls are stacked with maximum efficiency, then they are stacked to look like the following:

We can also roughly represent the stacking of balls like this:

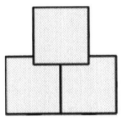

As we can see in this drawing, stacking the balls so they overlap allows us to stack more balls in a given unit of *height*. Rather than computing the volume of a ball, we need to compute the *effective* volume of a ball. That is, if we imagined little blocks around each ball, how much volume would that take up?

```
effective volume of ball
    = length * width * height
    = diameter of ball * diameter of ball * adjusted height
```

Step 6: Review & State Your Assumptions

We've got a lot of things to assume here! Let's update the prior equations with the values we'll assume.

```
volume of apartment
    = ceiling height x area of floor
    = 12 ft x 800 sq ft
```

```
= 9600 ft³
```

```
volume of furniture
    =   4 * volume of queen mattress
      + 1 * volume of couch
      + 1 * volume of loveseat
      + 2 * volume of dresser
    =   4 * 6 ft * 1 ft * 6 ft
      + 1 * 6 ft * 3 ft * 2 ft
      + 1 * 4 ft * 3 ft * 2 ft
      + 2 * 3 ft * 2 ft * 4 ft
    = 252 ft³
```

For the kitchen, we'll assume it's a 10 ft x 10 ft room, where the cabinets line two walls, going from the ground to the counter.

```
volume of appliances
    =   volume of sink & cabinet
      + volume of kitchen cabinets
    =   3 ft * 2 ft * 4 ft
      + 2 * 10 ft * 2 ft * 3 ft
    = 144 ft³
```

For personal items, we can reflect on moves that we've done. How many boxes (2 ft x 2 ft x 2 ft) of clothing and other items have we packed up?

```
volume of personal items
    = % items in closet * volume of personal items
    = 50% * 2 people * 10 boxes per person * 8 cu ft per box
    = 80 ft³
```

For the effective volume of a ball, we could do some more rigorous math to figure out what the adjusted height of the "cubed ball" will be. Or we can just look at the figure and eyeball it.

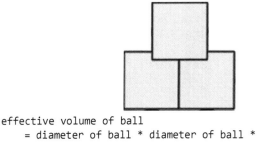

```
effective volume of ball
    = diameter of ball * diameter of ball * adjusted height
    = 2.5 in * 2.5 in * 2 in
    = 12 in³
    = 1/144 ft³
```

Careful! 12 cubic inches is *not* 1 cubic foot. It's 12/(12*12*12) cubic feet, or 1/144 ft³.

Step 7: Do the Math

We're basically all wrapped up—just a lot of arithmetic to do.

```
volume of apartment
    = 12 ft x 800 sq ft
    = 12 * 800 cu ft
    = 9600 ft³

volume of furniture = 252 ft³

volume of appliances = 144 cu ft

volume of personal items = 80 ft³

effective volume of ball = 1/144 cu ft.
```

Our final answer is:

```
usable volume
    =   volume of apartment
      - volume of furniture
      - volume of appliances
    = 9600 ft³ - 252 ft³ - 80 ft³
    = 9250 ft³


# balls
    = 9250 ft³ / (1/144 sq ft)
    = 1.4 million
```

We estimate that 1.4 million balls can fit in a two-bedroom apartment.

Step 8: Sanity Check

Does that 1.4 million balls sound high to you? It does sound a bit high to me. Let's do a quick spot check of our answer and ignore a lot of the details.

An 800 sq. ft. apartment with 12 ft. walls should be about 9600 sq. ft. If we can fit 1.4 million balls inside the apartment, then this means about 140 balls per cubic foot.

Does 140 balls per cubic foot seem high? Not so much. If we can fit about 5 balls along a foot-long side, then we can fit about 140 balls in a cubic foot.

Our math basically checks out. We'll give our interviewer the final answer of 1.4 million tennis balls.

Question 3: How many police officers are there in the US?

This one's going to get a little tricky, so let's get started. Remember: it's the approach that matters, not the final answer.

Step 1: Clarify the Question

The main thing that's ambiguous is who exactly we're counting as police officers. College campuses, for example, often have their own police officers. There's also some ambiguity over whether or not we're considering police officers who work "desk jobs."

Let's assume that we're referring to all police officers, including university police officers, whether they work a desk job or patrol.

Step 2: Catalog What You Know (or Wish You Knew)

We know:

- There are 300 million people in the US.

- More densely populated areas have more police officers.

- There must be police officers working all day and night.

- Police officers who patrol generally do so in pairs.

We might like to know the following:

- How much crime is there annually per capita?

- How many police officers work a desk job, relative to those on the street?

- How many police officers do you need per person?

The last of these, in particular, would be very nice. One way to get a *ballpark* estimate for the number of police officers needed is to use a specific example (a small town or school, for example) that we know of and work from there. Another way might be to ask what percent of people employed could be police officers. For example, it's unlikely that 1% of the US holds a job as a police officer.

Step 3: Make an Equation

One possible equation is something like this:

```
([population of the US] / [# of people per patrolling police
```

```
officer]) * (# of total police officers per patrolling officers)
```

Step 4: Think About Edge Cases and Sources

We've already covered the other sources for police officers (campuses, etc.), and ruled them out. We will have to think about how higher-crime vs. lower-crime areas will affect the police officer count.

Step 5: Break It Down

We know the population of the US. The tricky parts will be thinking about how many people there are per patrolling police officer and what the ratio of total police officers to patrolling officers is. We'll cover those in the next section though. In this problem, there's little to break down in the main equation.

Step 6: Review & State Your Assumptions

How can we get the number of people per patrolling police officer? A few ways, depending on what knowledge or assumptions you might have:

- I read an article about a particular school which had about 2000 students and 4 police officers. Schools, however, need a minimum staff to guard doors. An even larger school might have fewer police officers per person.

- A small town would need a minimum of 6 police officers (3 shifts per day with 2 officers per shift).

- Most police officers are male.

- At most, I could imagine that 1 out of every 200 men are police officers. This would mean about 1 out of every 400 people.

- There are probably more teachers than police officers. There must be at least 1 teacher for every school-aged child (between ages 6 and 18). Roughly 1/8th of the US population is between those ages, so that means that about 1 out of every 240 people is a teacher teaching kids between ages 6 and 18.

These assumptions offer a bunch of different numbers, but they're in the same general ballpark. We can probably take a guess at a number from there. Let's go with 1 patrolling police officer per 500 people.

Now, how can we guess the number of police officers per patrolling officer? Let's start with this: do you think that there's more desk work or patrolling work? My guess is that there's lots of paperwork for any incident. I'm going to take a guess at there being at least one person working a desk job for every officer out on the streets.

Step 7: Do the Math

We just have to plug our assumptions into the equation:

```
([population of the US] / [# of people per patrolling police
officer]) * (# of total police officers per patrolling officers)
=   (300 million people / 500 people per patrolling)
  * (2 officers per patrolling officer)
= 1.2 million police officers
```

Step 8: Sanity Check

How can we sanity check this? 1.2 million police officers means:

- About 1 out of every 300 people is a police officer. That's slightly less than the number of teachers of school-aged children.

- If the vast majority of police officers are male, then about 1 in every 150 men is a police officer.

- New York City should have about 26,000 police officers.

- San Francisco should have about 2500 police officers.

These all sound about right to me, except perhaps for the percent of men who are police officers. That seems a little high.

Actual Answer

There were about 861,000 police officers and detectives in 2006. The Bureau of Justice Statistics[3] offers the following additional facts:

- In 2008, local police departments had about 461,000 officers, accounting for about 60% of all state and local officers.

- Municipal and township police departments employed an average of 2.3 full-time officers per 1,000 residents.

Not bad! We got in the right ballpark.

Question 4: How many schools are there in the US?

Let's assume we're talking about public and private schools from kindergarten to 12th grade.

We can calculate this by estimating the number of students in public vs. private

---

3   "Local Police." Bureau of Labor Statistics. 10 November 2013.

school, and then the average size of each school.

Assume 300 million people in the US, an 80-year lifespan, and an even distribution across ages. This gives us about 50 million school-age children.

Number of Kids in Public vs. Private Schools: This is the tricky part. Let's divide the US population into lower, middle, and upper incomes. We'll conceptualize this as a pyramid, with 50% lower, 40% middle, and 10% upper.

- **Lower:** The bottom half of the US will generally not be able to afford private school, so 100% of them (roughly) are in public school. That's 25 million public school kids.

- **Middle:** This other 40% has a small number of kids in private school. Let's say 10%. This is about 2 million kids in private school and 18 million kids in public school.

- **Upper:** In the wealthiest areas (top 10%), I suspect about 20% of kids attend private school. That's 4 million kids in public school and 1 million kids in private school.

This gives us a total of 47 million public school kids and 3 million in private school.

Size of Public vs. Private Schools: We need to remember here the variance in public and private school sizes. Cities will tend to have bigger schools, but a lot of the US lives in smaller towns or suburbs.

- **Public Schools:** Let's assume that public schools are divided into sections of 4 grades each on average (elementary school, middle school, high school). The largest public schools might have as many as 1000 kids per grade, but there are also a lot of schools that are smaller (particularly in more rural areas). I suspect the average is more like 250 kids per grade, or 1000 kids per school.

- **Private Schools:** Private schools are more likely to have merged middle and high schools, so it's probably more like 5 grades per school. They won't (can't) vary as much in size. Private schools are more like 100 - 150 on the large side and 50 on the small side. Let's say an average of 75 kids per grade, so that's 375 kids per school.

Number of Schools: Now we just need to pull these together.

- **Public Schools:** We have 47 million kids and 1000 kids per school, so that's 47,000 public schools.

- **Private Schools:** 3 million private school kids at 375 kids per school is about

8000 private schools.

We have 55,000 schools in the US, of which about 15% are private schools.

Actual Answer

In 2008, there were about 133,000 schools in the US covering grades kindergarten through 12th grade in 2008[4]. Of those, about 99,000 (75%) were public schools.

55,000 million kids were in school and a little over 5 million attended private schools.

Not surprisingly, we were very accurate with the number of kids in school, and impressively close with the number of kids in private schools. We were a little less accurate on the number of total schools, but still fairly close.

Remember that it's the approach that matters, not the final answer. A problem like this is highly subject to your assumptions. If you assumed 500 kids per public school class, your answer would be much further from the actual number, even if you'd taken an identical approach. Fortunately, it's the approach that matters, not the actual end number.

Question 5: How long would it take to empty a hot tub using only a drinking straw?

I'll assume that someone is using the straw by filling it up and then emptying it on the side of the tub repeatedly (not, for example, using it as a hose with a continuous stream of water).

To estimate this, we'll need to calculate how big a typical hot tub is, what the volume of a typical straw is, and then the time to empty and refill a straw. We could then use the following equation:

```
([volume of hot tub] / [volume of straw]) * [time to empty and
fill straw]
```

This doesn't take evaporation of water into account. It also doesn't take into account that water at the bottom of the tub is harder to reach and therefore might take longer. We'll ignore these issues though, at least for now.

Volume of a Hot Tub: We can express the volume in cubic feet.

- **Depth of a Hot Tub:** Hot tubs have seats, which take up some space. What

---

4   "Digest of Education Statistics: 2010." National Center for Education Statistics. April 2011.

we really want to know is the average depth of a hot tub. If I picture a 6-foot person standing in a hot tub, the water comes a bit above their knees but well below their waist. That's probably about 2.5 feet deep from the water surface to the deepest point. When they sit though, it comes up about halfway to their chest. That is about 1.5 feet from the water surface to the seats. The floor is about half-covered with seats, so I think that gives us an average depth of around 2 feet.

- **Length of Side:** A hot tub can (cozily) fit about 3 people, each of whom are 2 feet wide. So a hot tub is probably about 6 feet wide.

The volume of a hot tub is therefore about 72 cubic feet (6 ft x 6 ft x 2 ft), or about 2 cubic meters.

Volume of a Straw: A drinking straw is probably about 20 cm long. I'd guess the width is about 0.5 cm long, which is a .25 cm radius. Let's round π (pi) down to 3, and convert again to metric, so we get a volume of 20 * (3 * .25²), or about 4 cubic cm.

Dividing 2 cubic meters (2,000,000 cubic cm) by 4 cubic cm gives us 500,000. So it takes about 500,000 trips with the straw to empty the hot tub.

Filling and Emptying a Straw: Just by "miming" this, it feels like this takes about 4 seconds.

We now have an answer: 500,000 straw trips * 4 seconds = 2,000,000 seconds.

Conversion to days: We can convert that to days if you'd like. 60 seconds / minute * 60 minutes / hour gives us 3600 seconds per hour. Multiply that by 24 hours (multiply by 10 to get 36000, then double to get 70000, then increase by about 25% to get about 90,000) and we get around 90,000 seconds in day. Dividing 2,000,000 by 90,000 will give us a little over 20 days. It's probably about 22 days.

Our final answer is 22 days.

Question 6: How many pairs of eyeglasses are sold every year in the US?

I'll assume that we're referring to eyeglasses only, not sunglasses.

There are about 300 million people in the US. Let's assume an 80-year lifespan on average. Let's also assume that people get new frames every three years on average. I now just need to calculate the number of people with eyeglasses in the US and divide that by three.

We'll need to break the population down by age since vision problems get worse with age. We'll also want to divide by gender, since (from my experience) a greater percent of men with poor vision wear glasses than women. Women are more likely to wear contact lenses.

There are also both nearsighted and farsighted individuals. Let's separate by this too.

Nearsighted: In my experience, very few young children are nearsighted. Moreover, those who are not nearsighted by age 20 or so tend to have stable vision for a long time.

Age	% of people of this age who are nearsighted	% of near-sighted men who wear glasses	% of near-sighted women who wear glasses	% of people this age with near-sighted glasses
< age 10	0%	0%	0%	= 0%
10 - 40	50%	50%	20%	= 17.5%
40 - 80	80%	80%	80%	= 64%

Weighting the column on the right by the proportion at each age, we get that about 38% of people are nearsighted with glasses.

Farsighted: Very few people become farsighted until well into adulthood. By around age 60, it seems nearly inevitable. Those who are farsighted only need correction part of the time and thus tend to wear glasses instead of contacts.

Age	% of people who are farsighted	% of farsighted men who wear glasses	% of farsighted women who wear glasses	% of people this age with near-sighted glasses
< age 10	0%	0%	0%	= 0%
10 - 40	0%	0%	0%	= 0%
40 - 60	50%	90%	90%	= 45%
60 - 80	80%	90%	90%	= 72%

This means that about 30% of people are farsighted with glasses.

Nearsighted or Farsighted: With 300 million people in the US, 30% of whom are farsighted with glasses and 38% of whom are nearsighted with glasses, this means that about 90 million people wear farsighted glasses and 114 million people wear nearsighted glasses. That's about 200 million people wearing glasses. (Note that this is actually double counting some people—some people will wear both nearsighted glasses and farsighted glasses—but that's okay for our purposes.)

If each person buys one pair of glasses every three years, then there are about 67 million pairs of glasses purchased each year.

Question 7: How much does a school bus weigh?

Let's assume we're talking about a school bus with a full tank of gas but no children on it. We'll also assume we're talking about a larger school bus, as opposed to a "short" bus.

How big is a school bus? A typical school bus has, as I recall, about 15 rows of seats. I think each seat is about 3 feet away from each other, so that's 45 feet, plus some extra room for the driver. Let's figure about 50 feet.

A bus is a little wider than a car, but can't be much wider or it wouldn't fit on the road. That makes it about six feet wide.

Let's break down the different components of a bus and calculate the weight of each. We have seats, a gas tank, tires, windows, an engine, and the frame.

Seats: There are 15 rows of seats, with 2 seats on each side, so that's 30 seats total. The seats are made from fairly sturdy metals, I believe, so they're not going to be super light. I'd guess they would be about 50 lbs. each. The seats therefore will be about 1500 lbs. total.

Gas Tank: I think my car holds about 20 gallons of gas, but a bus would probably have a much larger tank. Let's say it's three times larger. That's a 60 gallon tank. I think a gallon of water is around 10 lbs., so let's assume gas is the same weight. So that's 600 lbs. of gas.

Tires: A larger bus probably has three rows of tires, with two on each side, so that's 6 tires. I've lifted a car tire before and remember it wasn't too heavy. Let's say a car tire is about 20 lbs. A bus tire is larger though, so maybe 30 lbs. With 6 tires, that gives us about 180 lbs.

Bus Windows: Each window is about 2 ft. tall, I think. If we treat the left and right sides as having a continuous window (which is mostly correct), this is about 90 sq. ft. (45 ft. x 2 ft.) of window on each side. If we add in the windows at the front of the bus (about 6 ft. x 4 ft.) and the back of the bus (about 2 ft. x 6 ft.), this gets us an additional 36 ft. of windows. All in all, we're talking about 216 sq. ft. of windows.

How much does a window weigh though? Having lifted glass table tops before, I've found that a 3 ft. x 3 ft. piece of glass is fairly easily lifted, but still heavy. I'd guess that that's about 40 lbs. This gives us an estimate of about 4 lbs. per sq. ft.

So 216 sq. ft. of windows at 4 lbs. per sq. ft. would be about 850 lbs.

Engine: This is tricky to guess since I don't have much experience with engines. I'd guess though that you'd need two or three strong people to lift a car engine. If each person can lift 100 lbs., that gives us a guess of about 300 lbs. Again, a bus will have a substantially larger engine, so let's say about 500 lbs. for a bus engine.

Bus Frame: This is probably the hardest part as I don't really know how much metal weighs. We have a metal frame below, which is about 45 ft. long by 6 ft. wide. This needs to be fairly thick. I would guess that 1 sq. ft. of the frame is about 30 lbs., so that's about 8000 lbs. for the bottom.

The top doesn't need to be quite so thick, so let's say about 4000 lbs.

Each side of the bus is about 6 ft. tall and 45 ft. long. If we go with an estimate of 15 lbs. per sq. ft., that's about 4000 lbs. per side, or 8000 lbs. across the two sides.

The front and back of the bus is 6 ft. tall and 6 ft. wide. I'd guess we need fairly sturdy metal there, so let's again assume 30 lbs. per square foot. That's about 1000 lbs. for the front and the back, or 2000 lbs. together.

Our entire frame is then about 22,000 lbs.

Total: We have a 22,000 lbs. frame, 500 lbs. for the engine, 850 lbs. of windows, 180 lbs. of tires, 600 lbs. of gas, and 1500 lbs. of seats. In total, that's 25,630 lbs. I think a car is about 3000 lbs., so 25,000 lbs. for a school bus seems like it's in the right ballpark.

Actual Answer: Depending on the size, a school bus can be anywhere from 10,000 lbs. to 36,000 lbs. A typical 62-passenger bus (about what we assumed here) is around 20,000 lbs., without gas.

Question 8: How many basketballs are purchased every year in the US?

Basketballs are purchased by schools, (non-school) teams, and families. Let's calculate how many basketballs are owned by each group and then estimate a replacement rate.

I'll assume that basketballs purchased by adults for their own use (not for their kids) is negligible.

Families: There are about 45 million kids aged 6 - 18 in the US. Let's assume that families without kids are unlikely to own basketballs, and each family has an average of 2 kids. This means that there are about 20 million households with kids.

I'd guess that one fourth of households with kids have a basketball, and they buy one basketball per year on average. This is about 5 million basketballs per year.

Schools: My high school had about 50 basketballs for a 500-person school. With constant usage, each ball probably didn't last more than three months. There are nine months in the school year, so this means we went through about 150 basketballs per year for 500 kids.

However, my school was better funded than most schools in the US. A ratio of 50 basketballs per 500 high schoolers might be more realistic.

We should actually adjust our number downwards even more since elementary schools won't use as many basketballs. Let's figure then about 25 basketballs for every 500 kids, or 1 basketball for every 20 kids.

With 45 million kids, this means about 2 million basketballs purchased by schools every year.

Teams: There are a variety of teams in the US: kids teams, professional teams, adult intramural teams, college teams, and school teams. We've already included school purchases, so we don't need to recount those.

- Kids teams: If we assume that 1% of kids play on a basketball team and a basketball team has about 10 kids per team, then there are about 45,000 kids' basketball teams in the US. If each team buys 10 basketballs per year (one per player), then this is about 450,000 basketballs.

- Professional teams: This should be negligible.

- Adult intramural teams: This should be negligible.

- College teams: There are about 15 million college-age people in the US. Of those, maybe 5% play a college sport, and for 5% of those people that sport is basketball. Thus there are about 37,500 college basketball players in the US. If we again assume a purchase of one basketball per player per year, this is 37,500 basketballs.

This gives us about 500,000 basketballs purchased for teams.

We have accounted for 5 million home purchases, 2 million school purchases, and 500,000 team purchases. This gives us a total of about 7.5 million basketballs purchased every year.

Question 9: How much money do people spend on haircuts every year in the US?

Observe first that the price of a haircut can vary dramatically. A haircut, plus dyeing of hair (common for women), can be $200 or more at a nice salon. On the other hand, a man's haircut at a cheaper barber might be as low as $10. We'll need to consider this in our answer.

We can assume that men get their hair cut 12 times per year (once per month) and women get their hair cut about 5 times per year.

I'd guess that about two-thirds of people (adults and children) get their hair cut professionally, with the remainder getting their hair cut at home or through a friend. This gives us about 100 million men and 100 million women using professional hair cutters.

Therefore, we have about 1.2 billion male haircuts and 500 million female haircuts per year.

I rarely see men in nice salons, so we can assume that virtually all of the 1.2 billion male haircuts take place at a cheaper haircutting place. If the average price of a haircut there is $20, then we have about $24 billion spent by men on haircuts.

Many more women go to nicer salons for their haircut. Let's suppose that 20% of women go to nice salons for their haircut and that those haircuts cost an average of $100 each. This means that the average woman's haircut costs $36. Women therefore spend about $18 billion on haircuts every year.

Combined, that's about $40 billion on haircuts each year. However, I think that was overestimating a bit. I suspect about 25% of the US doesn't need haircuts because they're bald, small children, and so on. The number is probably closer to $25 million.

Question 10: How much money does Facebook make in ads every year?

I remember seeing reports not too long ago that Facebook has about one billion users. However, many of those accounts might not be active. Let's assume that 50% of those accounts are actually active and the active users log into Facebook about once per day on average.

This means that Facebook has about 500 million visits per day. The average visit is probably about 10 minutes and has about 10 pageviews. This gives us a figure of 5 billion pageviews per day.

If each page has four ads on it, then Facebook shows about 20 billion ads every day.

As I recall, click-through rates for search ads on Google are usually about 2%, but I also know that search ads get substantially more clicks than display ads. If we assume a 10x difference between search ads and display ads, then we have a click-through rate of about 0.2%.

This gives us a total of about 40 million clicks per day.

How much revenue does each click generate?

I know Google search ads rarely generate less than 5 cents per click but they can generate as much as $10 per click. This is a huge range, of course. From my experiments with ads, I found that I was paying about 25 cents per click. This was probably for a slightly cheaper-than-average market segment though. Let's figure about 50 cents per click on average. I'm assuming here that Facebook ads are equivalent to Google ads, which could possibly be an unfair assumption. Let's work with that for now though.

40 million clicks per day at 50 cents per click is about $20 million per day. This means about $7 billion dollars in revenue.

This sounds about right if we look at salary costs. If Facebook has about 5000 employees, each of whom probably earns about $100k in salary on average, then this is $500 million in salary costs. It seems reasonable to think that Facebook's revenue is 10 - 20x its salary costs.

Actual Answer: Facebook's annual revenue was $5.1 billion in 2012, about $4.3 billion of which was from ads.

Product Questions
Chapter 14

When Jessica interviewed at Apple for a role with iTunes, she was asked a simple, benign question: "What do you think of iTunes? What would you change?" She was well prepared for this question, as she loved music.

She immediately began listing a bunch of features she would implement or change. Better keyboard navigation. Easier ability to cancel downloads. Better search in the iTunes store. The list went on and on.

She had a lot of opinions and her ideas were, on the whole, fairly good. When she didn't get the next interview, she figured the interviewer was hurt by her being too harsh.

Not exactly.

It's okay to love or to hate a product, but you need to do it the right way.

About the Product Question

The product question is the heart and soul of the PM interview. It directly gets at what a PM does: design, build, and improve products.

These questions come in three common forms:

1. How would you design _____? For example, you might be asked to design an alarm clock for the blind.

2. What would you improve about _____?

3. What's your favorite product and why?

While these questions sound different, they have one very important aspect in common: You need to understand and focus on the goal. The goal might be to design the best product for the user or it might be to increase revenue or another metric.

Type 1: Designing a Product

These questions aren't nearly as open-ended as they sound. You want to approach these questions like a good PM would: with a structured approach that starts with the user.

Remember: it's not about what *you* want the product to be. It's about what the user wants.

The Approach

We've offered one framework that works well for these problems, but there are other frameworks too. Good frameworks have the following in common: they ask appropriate questions, understand and assess a goal (often a good user experience), and apply a structured approach to accomplish that goal.

Step 1: Ask questions to understand the problem

Before you can even start to answer the question, you need to make sure you understand what the question is. It might not be what you think.

For example, suppose you were asked, "Design a pen." That's a pretty straightforward question, right? Not necessarily.

The pen could be:

- A permanent marker, designed to not come off in the laundry.

- A pen that uses ink that only shows up under special lighting.

- A pen for astronauts to use in space.

- A pen for children to use in the bathtub.

- A pen for scuba divers to use.

Clearly, each of those people would need a very different pen. They would need a different size, color, and feature set.

Are they trying to trick you? Yep, in a sense. However, a PM who just dives into creating a product without understanding the goals might create something that is radically different from what the user needs.

Step 2: Provide a structure

Interviewers are looking for structured thinking. The easiest way to show this is to give a structured answer and call out which part of the structure you're on. For example, you might say something like, "First I'm going to talk about the goals. Then, I'm going to list out some potential features. Finally, I'm going to evaluate each of those features against the goals. Okay, so starting with the goals...".

This will communicate to your interviewer that you approach problems in a structured way. It will also help keep your thoughts and those of your interviewer straight.

Step 3: Identify the users and customers

Now that you understand the question itself, you should identify who the users and customers are. Ask more questions if you need to.

In some cases, the users and customers aren't the same person. The customer is the person paying for the product; the user is the one using it. There also may be multiple users.

Example: "Design a calculator for kids"

In this case, the interviewer has told us who the user is. Or have they? There are, of course, many kids in the world and they're not all the same. The child is also not the only user.

We have the following potential users or customers:

- **The child:** The child is the primary user of the calculator. You will need to know children's ages. A calculator for seven year olds will look very different from a calculator for high schoolers.

- **The teacher:** If this is a calculator used in schools or for an academic purpose, then the child's teacher will likely need to use the calculator or at least understand how it works.

- **The parent:** The parent may use the calculator when helping a child with homework, and they are likely paying for the calculator as well. This makes the parent both a user and the customer.

Depending on the type of calculator, there could be even more users. For example, if the calculator is designed specifically for a textbook, then you might also include textbook publishers. Or, if the calculator is being used within a classroom, the school or school district might be the purchaser. This might require special consideration.

Example: "Design a better stove"

As with the "design a pen" example, the stove could be a bit different than we imagine. Is it for household usage? For a large restaurant? For kids? We'll need to ask questions to figure this out.

Let's suppose there's nothing funky going on here. We're just designing a stove for normal home usage. Still, we'll want to think about the different types of users. This could include:

- **Novice Cooks:** These people are new to cooking and may want something simpler.

- **Advanced Cooks:** These people are more advanced and might want a bunch of advanced settings.

- **Children:** Even if children aren't using our stove, they're still around a household. A good household stove can't pose a safety danger to children.

- **Elderly or Disabled:** Elderly or disabled people might have slightly different requirements since their physical mobility is limited. They might also have special meals or cooking needs.

Each of these people will have different requirements.

Thinking about Users

With each question, think about where the product is being used and who else might interact with it. As a good rule of thumb:

- Children's products may be used by children, their parents, and their teachers. The parent or the school might be the customer.

- Healthcare products (including products for people with disabilities) might be used by patients, doctors, and insurance companies.

- Sports-related products might be used by athletes and their coaches.

- Products for a professional (e.g., accounting software) might used by the

---

professional, her assistant, and others in her company.

We have to design for all of those people, so it's important to call out who they are.

Step 4: What are the use cases? Why are they using this product? What are their goals?

For each user (if there's more than one), make a list of the use cases. This is a list of the different tasks or scenarios that a user might want to use the product for.

For example, if we're designing a keychain for the elderly, the use cases might include:

- Locating the right key in the keychain to open their house, enter their car, etc.

- Adding a new key to the keychain.

- Removing a key from the keychain.

- Finding the keys in the bag.

- Finding the keys in their house.

You'll need to assess, either by yourself or by discussing the situation with your interviewer, which use cases to design for. You might decide that all of them are very important, or you might decide that some use cases are less important than others.

You can also think about these goals at a higher level. You can think about not only *what* you do with a product, but *why you do it*. What is the underlying motivation? For example, the underlying motivation for the keychain might be independence.

You want to convince your interviewer that your product will change the world by appealing to the underlying motivations, goals, and use cases.

Step 5: How well is the current product doing for their use cases? Are there obvious weak spots?

Go through each use case and assess how well the current products or solutions address those. What are the user's biggest issues with the product? These are the areas you will focus your design on.

If there are multiple users (for example, the elderly person and their caretaker), we may need to assess their use cases separately.

In many cases, and especially when you get a question in the form of "Design a
_____ for the _____," it can be useful to think carefully about what makes
this type of user special. For example, an elderly person often has limited
mobility and dexterity, but they are about more than just their limitations; they
also have particular values. They might care deeply about family connections, or
prioritize healthcare or stability. We'll need to keep this in mind for our design.

Step 6: What features or changes would improve those weak spots?

Up until now, we've just been assessing the current problem and needs. It's
good we spent all that time defining the problem. This will help us come up with
a solution that's truly tailored to their needs, rather than to what you personally
would want.

In many cases, we will want to solve the issues with multiple use cases at once.
For example, the solution to adding a key to the keychain is very closely tied to
the solution to remove a key from the keychain.

A good way to handle this section is to name a few ideas and then ask the interviewer if they want you to dive deeper into any of them.

Make sure to explicitly tie your feature ideas to the use cases or goals. Make it
really, really clear you're coming up with ideas that are customer focused, not
just things you've always wanted.

If you start to run out of ideas, go back to your use cases and be willing to get a
little more creative. If the interviewer really seems to expect you to have more
ideas, ask if there are any that she thinks you didn't explore sufficiently.

This is a good spot to use the whiteboard as well.

Step 7: Wrap things up

As a final step in the interview, it can be useful to give the interviewer an
overview of your solution. This is especially important if you've been talking for
a while. You might have gone over many solutions to the problem, and your
interviewer may be unclear as to your current proposal.

If you haven't touched the whiteboard yet, this may be a good time to do so.

Example: Design an Alarm Clock for the Blind

Let's walk through this problem step by step. Note that this is only *one* solution.
There are many ways to respond to this question.

Step 1: Ask questions to understand the problem

In this situation, we're told explicitly who the user is. However, this information may be only half true.

What kind of blind person? Many people who are blind can detect light or even see blurry shapes. Second, blind people can be children, adults, or elderly people. They may even have additional disabilities. If we're designing for a specific type of blind person, this will affect our product.

We should also understand where the blind person is using the alarm clock. Will this be an alarm used at home, or perhaps one for travel? Is it even a physical alarm clock, or could the interviewer be asking about a mobile app designed for blind people?

Let's assume our interviewer confirms he's thinking of a fully blind person—zero ability to see—and that he would like us to design for a blind adult who will be using a physical alarm clock at home.

Step 2: Provide a structure

The approach in this problem is essentially the structure we offer to the interviewer.

> Okay, now that I understand the problem, I'm going to tackle this in a few parts. First, I'll think about who the users are and what they're using the alarm clock for. Second, I'm going to compare existing alarm clocks against these use cases to see where the gaps are. Then finally, I'm going to discuss how we can fill in these gaps.

As you go through these steps, make it clear to your interviewer when you're

Phrasing Questions

Be careful about how you phrase your questions. A question like, "What does the blind person do?" can sound like you want your interviewer to solve problems for you. However, if you word it as "Is there anything we know about what a blind person does?" makes it clear you're trying to work collaboratively with your interviewer.

If your interviewer pushes back with a question such as "What do you think?" that's a sign to stop pushing with the questions.

transitioning from one step to the next. You could say something like, "Now that we've identified the users, let's move on to evaluating existing alarm clocks."

Step 3: Identify the users and customers

We have one user in mind already: a fully blind adult.

Who else might use the alarm clock? A blind person is likely around non-blind people, such as their spouse, children, or healthcare workers. Could you imagine if your spouse bought an alarm clock that you couldn't turn off? Yikes! An alarm clock for the blind still needs to be moderately usable by non-blind people.

Step 4: What are the use cases? Why are they using this product?

The core use case here is presumably to wake up from sleep, usually in the morning but possibly for naps as well. This means he'll need to check if the alarm is set, configure the time, enable the alarm, be woken up by it, snooze the alarm, and disable it.

Could there be other reasons why the blind person uses the product? Sure!

- He could be using it to check the current time.

- He could be using it to time something (e.g., boiling an egg). After all, not all his appliances will be adapted for his disability, so he might use the alarm clock to compensate.

- He could be using it to remind himself to take medicines. In this case, he'll probably need some sort of recurring alarm.

- He may be using it to listen to music or the radio, as many people do with their alarm clocks.

I'll assume the primary function we want to design for is the wake-up-from-sleep one.

Checking the current time is certainly an important use case for alarm clocks in general, but it may not be an essential one for blind people. People use an alarm clock to check the time because it's highly convenient (that is, the time is right there, staring at them from across the room). Unless we can achieve similar convenience, blind people probably wouldn't use the alarm clock for this purpose. They would default to using whatever they otherwise use to check the time.

Step 5: How well is the current product doing for their use cases? Are there obvious weak spots?

A standard alarm clock relies on visual indicators for almost everything. We know if the alarm is on or off based on a light on the alarm clock. We know the current time because we can read it on the display. We just look at the alarm clock to know if the device is plugged in. Essentially, the only thing we don't use a visual indicator for is waking up.

This poses a number of challenges for a blind person, since he won't be able to see the visual indicator.

- **Enabling / disabling the alarm:** Alarms commonly have a single button to toggle the alarm; press the "alarm set" button to flip the state. This is fine when you can see if the alarm is set, but it's likely to be problematic for a blind person. Toggling the state of something only works well if the current state is very obvious.

- **Ensuring the alarm is plugged in:** Electronics can get unplugged for any number of reasons. Simply giving the blind person a way to check if the alarm is plugged in might not be enough, since they probably won't remember to check this option before going to bed. Rather, we need to make it *obvious* that the alarm is unplugged.

- **Checking if the alarm is set:** A standard alarm clock has a light on the display to indicate whether the alarm is enabled. This won't work for blind people since they can't see the indicator light.

- **Setting the alarm:** A standard alarm clock is set by pushing an "up" or "down" button on the time, and waiting until the time is set correctly. Since a blind person can't read the clock, the visual feedback on the time doesn't really help them.

- **Setting / checking the current time:** Although we don't need to use the alarm clock to check the time, we do need to ensure the current time is set correctly. Since a standard alarm clock relies on a visual display for the time, this doesn't work well for blind people.

We now need to resolve these issues.

Step 6: What features would improve those weak spots?

We want to keep the alarm clock simple to use. Complicated designs are no fun for anyone, blind or not.

Design 1: Audio Playback

The major issue we'll need to design around is the lack of visual display. As a very simple approach, we can use audio playback, in addition to a visual display.

The benefit here is it's easy to build audio playback, and the use of a visual display would make the design usable for non-blind people as well.

However, it has a major drawback: it might wake up the blind person's partner or spouse. We'll work with this approach for now, though.

Turning on / off alarm + Ensuring the alarm is plugged in

Like a standard alarm clock, this alarm clock could have a button that turns the alarm on and off. We will provide audio feedback when we set the alarm to indicate either "the alarm is on" or "the alarm is off." This also makes it easy to ensure the alarm didn't get secretly unplugged.

Checking if the alarm is set

To keep things simple, we can just reuse the button that turns the alarm on and off. We don't really need a special button for this.

Snoozing the alarm

To snooze, we will have a separate "snooze" button, much like a normal alarm.

Setting the alarm time

To set the current alarm, we can offer "up" and "down" buttons to change the hour and minute hands. When we press a button, it announces the current time.

Doing this on every single button press might be a bit slow though, particularly when changing the minute. Instead, we can design it so that if you hold down the "up" or "down" buttons, it suppresses the audio playback until the button is released.

Setting the current time

We can set the current time essentially the same way that we would set the alarm time.

Design 2: Limited Audio Feedback

As we said earlier, the issue with using audio playback is that it might disturb others in the room. We may want to design a more advanced alarm clock that doesn't rely on audio playback for the most common functions (particularly those that might be done while the user's partner is asleep): turning on and off the alarm and setting the alarm time.

Turning on and off the alarm

Instead of audio feedback, we could use a physical switch (much like a light switch) to turn the alarm clock on and off. The risk here is that the alarm clock could wind up unplugged and the user wouldn't know.

Instead, we can still use a single button to toggle the alarm. Instead of audio feedback, the clock will vibrate once to say "alarm is on" and twice to say "alarm is off."

Checking if the alarm is set

Again, we will just reuse the button which turns the alarm on and off. The vibration feedback will tell us if the alarm is set.

Setting the alarm time

To set the alarm time, we can have several rows of buttons on the side of the alarm clock. Each button would have the number written in braille as well as in colored text for a non-blind person. Since we're using buttons to set the time, we don't need to use audio playback.

Hour	12	1	2	3	4	5	6	7	8	9	10	11
Minute	0	1	2	3	4	5	6	7	8	9	AM / PM	
	0	1	2	3	4	5	6	7	8	9		

The AM / PM button would be pushed in to indicate it's set to AM, otherwise it would be set to PM.

This might be far more buttons than we really need. Does anyone really need their alarm clock to be capable of setting their alarm for exactly 9:23 a.m.? Probably not. Setting the alarm to 9:25 or 9:20 would work just fine.

Let's remove some of these buttons to clean up our design. Simplicity is a good thing in design.

Hour		
12	1	2
3	4	5
6	7	9
9	10	11
AM / PM		

Minute		
:00	:05	:10
:15	:20	:25
:30	:35	:40
:45	:50	:55

We could potentially simplify this even more, by just listing :00, :15, :30, and :45.

Alternatively, we could replace the minute buttons with a circular dial. However, a dial for the hours could be a bad idea, since someone might accidentally set the switch to the wrong hour.

Step 7: Wrap it up

Depending on the priority we place on not disturbing the user's partner, we may or may not use audio feedback. You could mention to your interviewer the sales benefit of marketing the alarm clock as non-intrusive to bed partners.

The first design uses audio feedback to basically replicate what the physical display does. We still have a visual display, in case non-blind people use the device, but we announce any alarm or other changes out loud. The only non-obvious change here is suppressing the time-change announcement when we hold down the "up" or "down" arrows.

If we're concerned about waking up the user's partner, we can remove the audio feedback and replace it with vibration (to indicate the alarm status) and physical buttons (to set the time).

In either case, we will keep the physical display to ensure the device is usable by non-blind people as well.

Type 2: Improving a Product

These questions can be asked in general terms ("Pick a product. How would you improve it?") or by targeting a specific product ("How would you improve Product X?"). Either way, the secret to these questions is identifying and understanding the product's biggest issues.

You may notice the approach for this type of problem will be very similar to "Design a product" questions. This is to be expected. The key difference is you're taking an existing product, assessing the issues, and improving it from there.

Structure is, again, very important here. You don't have to be quite as explicit in outlining your structure, but you should tackle it in an organized fashion. A simple line like this will work: "Let me start first with understanding the goals of the product, then move on to the issues and how to solve those. Okay, so the goals of the product are…"

Step 1: What is the goal of the product?

First, you need to understand the product's ultimate goal. What problems is it solving for the user?

Facebook, for example, allows you to post status updates and share photos. That's not the *goal* of the product, though. The goal is to connect and share your life with your family and friends.

Note that a product might have primary and secondary goals. Facebook's primary goal for consumers is connecting with family and friends, but business users have a different goal: engaging with existing and potential customers.

Step 2: What problems does the product face?

Next, you need to assess the problems the product faces.

- Does it need to expand its user base? If so, should it broaden its user base by entering a new market, or should it expand in its existing market?

- Does it need to increase revenue? If so, is this about increasing revenue per user or about increasing the number of paying users?

- Does it need to increase user engagement?

- Does it need to increase conversions from visitors to registered users?

To assess this, think about how the product is currently designed. What does the product appear to prioritize?

Of course, you can also ask your interviewer what the key problems are, but you don't want to come off as being unable to figure this out on your own. You might therefore want to make some educated guesses about what area to focus your improvements on, and then validate this direction with the interviewer.

Step 3: How would you solve this problem?

Third, brainstorm a few ways you might solve this problem and discuss the pros and cons of each. Again, be open about the tradeoffs of each option.

Your ideas might be bold, crazy ideas, or they might be small, iterative improvements. Both approaches (and everything in between) have immense value in the real world.

However, some companies have more of a preference for one or the other based on their size, risk tolerance, budget, or culture. If you know which way the company leans, you can focus your discussion there. Otherwise, preface what you're suggesting accordingly:

> We can make a few quick fixes that will help mitigate this issue. However, if we are willing to take a bigger gamble, some additional options are open to us.
>
> As far as the quick fixes, we can ...

Doing this will allow you to demonstrate you understand that while big changes come with big rewards, they also come with big risks. A good PM needs to balance those.

Step 4: How would you implement these solutions?

If it's not immediately obvious, discuss with your interviewer how you would implement the solution you proposed. What are the bigger technical or business challenges? How could you reduce the costs or risks associated with the solution? For example, you might test out your solution on a small user base or roll out a limited prototype.

Many startups in particular will value a PM who is "scrappy" and can do a lot with limited resources. Show this side of yourself.

Step 5: How would you validate your solution?

A good PM knows he's not always right and that his suggestions are little more than an educated guess. Therefore, he will prioritize validating his solution.

Discuss with your interviewer what metrics you would gather to see if your solution really worked. For example, if you suggest sending "People who bought X also bought Y" emails to users in order to increase revenue per user, you may want to track metrics such as email clicks, direct referrals to product Y, future purchases of product Y from this set of users, and purchases of other products. You might also want to come up with some approaches to ensure the emails don't harass users or make them unsubscribe from all emails.

Type 3: Favorite Product

It's very likely that at least one of your interviewers will ask you what your favorite product is and why (see our preparation tips below).

These problems are similar to "improve a product," but approached in reverse. Rather than speaking about what's broken about a product, you discuss why you love this product.

Generally, people will speak about a product they personally use. However, you could pick something you don't use but love anyway. For example, a parent could think a child's toy is brilliantly done and thus their "favorite" product, even though it's not one that he or she personally uses.

Some of your favorite products might not be good candidates for this question; beware of discussing them if you just can't think of much that's interesting to say about them.

Similarly, you shouldn't pick a product solely because it's an interesting one to discuss. You want something that you, personally, connect with. You want your passion for this product to come out. Don't pick something because it's the "right" product if it's not the right one for you.

As always, you want to structure your answer to these questions. This question doesn't necessarily require you to explicitly state your structure (although you could, if it sounded natural), but you should still keep your response organized. Strong communication skills are always important.

Step 1: What problems does the product solve for the user?

You, or at least someone close to you, are probably the user here. You need to think about not just what the product is, but also what the user's goals are.

For example, take a product that analyzes your personal finances, like Mint.com. That's what the product does, but it's not a goal. The goal is likely to help you save money by understanding how you're spending it. Did you really want to spend $1,000 on food last month?

You can discuss more than one goal, but don't go overboard. If you're listing more than one or two goals, your discussion will lose focus—and you might actually be discussing features, not goals.

Step 2: How does the product accomplish these goals? What makes it "neat"? What makes users fall in love with the product?

Now you should discuss how exactly the product accomplishes these goals. There is something about the product that makes it uniquely powerful at doing this.

For example, it might have a wealth of user data that allows it to perform more powerful analytics. Or maybe it just has an excellent interface, where every button seems to be in just the right place.

In addition to being great at accomplishing these goals, there might be other things about the business that you think are particularly impressive. They might have a clever revenue model that's a win-win for users, or they might have a way of reducing their costs substantially by relying on crowdsourcing. It's okay to talk about these details as well.

For many products, there is also an emotional connection to the product. Some products are truly loved by consumers; they are the things people rave about. Why?

Step 3: How does it compare to the alternatives?

Every product has its alternatives, even ones that are seemingly first to market. Think both narrowly and broadly about what the "competition" is for this product. The direct competition might be other websites that do the exact same thing, and the indirect competition might be physical products that achieve your goal.

In other words, there is a reason why you're not using the alternatives. What is that reason?

Step 4: How would you improve it?

You don't necessarily have to go into this immediately, but it's a natural follow-up question the interviewer might ask. You should approach this the same way you would with the prior type of question.

Example: What's your favorite website?

Here's one answer I personally might give:

> "I'm a big fan of Quora, which is a question-and-answer website where users can pose questions and have them answered—often by experts in the subject. For example, someone might ask, 'What's the best way

to replace a flat tire?' or 'Should start-ups wait to raise venture capital?', or even 'What is it like to live on a farm?' It's everything from solve-my-problem questions to what's-your-opinion-on questions to tell-me-your-experience questions.

What I love about Quora is how effective it is for solving specific problems. There are experts in virtually every industry offering their expertise and giving extremely thorough, well-thought-out answers.

It's helped me on a bunch of occasions. For example, when I was working on a translated version of some content, I needed to know how a Japanese translation would compare in number of pages to an English version. I got several great answers, approaching it in different ways. These people all really knew what they were talking about.

On another occasion, I just wanted to know where the best coffee shop that would be good for studying near my home was. A lot of local people on the site and I found my now-favorite cafe this way.

But even beyond solving specific problems, I find I just love being on the site. It's essentially a community built on learning and exploring. You get to know the people on the site and value their responses, because you can trust them. I've never been the type of person to make "friends" through Quora. And yet, on Quora, I really do feel like I know the people. It's a real community.

I think it's one of very few sites which have been able to bridge that gap between learning and fun. It's engaging. Because it's question-and-answer style, I find myself stumbling across a topic I'd never been interested in but wondering, "Yes, why is that?"

In many ways, it's replaced Google for me as a place to get questions answered. If you think about it, searching online to solve questions has several issues:

» *First, you search by keywords, even when you're trying to answer a question. Your hope is you'll find a page that offers a broad enough overview that you can answer your question. It's not truly solution-oriented, so it's often inefficient.*

» *Second, trust. You might not trust the webpage author's credentials. And even if you do, authors are wrong even about their areas of expertise. Quora allows users to give feedback on answers via comments (as well as through upvoting / downvoting), which is a great way to validate an answer's accuracy. Google's version of reputation is through PageRank, which is pretty imperfect.*

> » *Third, a lot of information is just hard to access through normal webpages and blogs. Websites have an inherent bias towards authors with the capability and desire to create a webpage. It's easy to find technical support, but what about tips on how to act out a specific part from a play? Not as easy.*
>
> *Of course, I still search for answers online, but I find myself using Quora more and more both to solve my problems and as a place to browse and learn new things."*

I've critiqued Quora here, but I've also done a few more things.

- I've implicitly demonstrated a love for learning.

- I've discussed why Quora grabs me: it's question-based and it's a community. This speaks to the emotional attachment some people have to a product.

- I've discussed some less obvious flaws of online searching.

This isn't necessarily the answer for you, of course. Think about a website or product that speaks to you.

Preparation

Practice makes perfect—and you *should* practice these questions alone or with a friend. Additionally, it's important to come in with ready-to-go answers for these questions.

Step 1: Select Products

Walking into your interview, you should be prepared to talk in depth about the following:

- One online product.

- One physical "offline" product. Interviewers love to try to "stump" candidates by asking them to assess a physical product.

- One product you purchased recently.

- Your "favorite" product or website.

- A product you think is well designed.

- The company or team's product.

Oftentimes, a company will ask a candidate "What's the last product you purchased?" It doesn't truly need to be the very last product, but it should be something fairly recent.

Some of these products could be the same. For example, the online product you selected could be your favorite product. However, being prepared to discuss more products may prove valuable.

Step 2: Understand Key Metrics

For each product, think about and understand the product's key metrics. These will likely include the following:

- **Users / Traffic:** How many users does the product have? How are they acquiring users?

- **Conversion:** How effectively does the product convert a visitor to a user, a free user to a paid user, or a paid user to a more highly paying user?

- **Referral Rates:** Do users refer other users? How often? Is the product viral?

- **Engagement:** Are users actively engaging with the product (posting, comment, playing, etc.)? How often?

- **Retention:** How often do users come back? How many users come back after a certain amount of time? For some products, visiting once a month is good. For others, you'll want a user to frequently visit.

- **Revenue:** How does the product make money? How much money does it make?

- **Costs:** Where does the product face costs? Does it require physical materials? What are its development costs? Does it face high support or sales costs?

It's important to think through which of these are most crucial, which the product excels in, and which it struggles in. What would you do to change these metrics?

See Also: Product Metrics, pg 252

Step 3: Analyze Each Product

Now that you understand what the metrics are for each product (as well as you can), analyze the product on the following aspects:

- **Users and Goals:** What are the primary types of users? What are their goals? How does the product help users accomplish these goals?

- **Strengths:** For which metrics does the product excel? Does it have a lot of users? Does it have high engagement per user?

- **Challenges / Focuses:** What is the main challenge the product faces? Is it struggling to get users to sign up? Is it struggling to convert free users to paid users?

- **Why, Why, Why:** Why does the product excel (or struggle) in a particular way?

- **Priorities and Values:** What does the product or company care about? Apple, for instance, is obsessed with the user experience. An enterprise-software company, however, might prioritize security and reliability over aesthetics.

- **Competitors:** What are the product's competitors? How does the product measure up to them? Think both narrowly and broadly about competitors. A stereo competes not only with other stereos, but also with people's mobile phones.

- **Tradeoffs:** What are the tradeoffs a product has had to make in order to accomplish its goals or address user needs? How did it come to be that the product had these strengths or challenges?

In essence, you want to dissect the product from a business perspective, as well as from a user perspective. Keep asking yourself *why* and *how*. Why did the product do something in a particular way, and how could it be better?

While you should assume the company probably had good reasons for its decisions, you shouldn't automatically assume that they did everything right. Virtually every product can be improved.

Tips and Tricks

Ultimately, these design questions are getting at how well you can show user empathy. Can you get inside the user's head and think about what that type of user would want? Or will you just design the product you will want? Focusing on the user is key.

In addition though, the following tips and tricks will help you:

- **Have an Opinion:** Develop a point of view and act like you're the owner of the product. Interviewers want PMs who have opinions.

- **"Wow" the Interviewer:** Try to come up with at least one "wow" idea. When you find it, point it out. Did you think of a major market the product could enter easily? Did you come up with a killer feature that would make people use the product twice as much? Don't let it get hidden among the quick fixes.

- **Use the whiteboard:** Don't feel you need to stay glued to your seat during such a question. In fact, it's great to get up and use the whiteboard; that's what it's there for. Using the whiteboard may help you communicate your design more clearly to your interviewer.

- **Don't overbuild:** Some people go overboard with wacky designs to solve a simple problem. You'll want to be realistic about what can and can't be built, and what the tradeoffs are.

- **Don't complain that you need more research:** Yes, yes, we understand that, in an ideal world, you'd love to study a variety of things and gather a bunch of data. You can't do that in the interview though—or in the real world. You'll have to make do with what you can. It's fine to briefly discuss what research you would like to gather to better inform your decision. If your interviewer seems interested in this discussion, then go for it. Otherwise, don't dwell on it. Use your best judgement to inform your decision, just as you would in the real world.

- **Think about the business, too:** The customer's goals are incredibly important, but so are the goals of the business. If you're critiquing a real product, think about the business goals. For example, if you're discussing what you would improve about a social media product, consider what the business' primary concerns are. Is it increasing revenue? User sign ups? User engagement? These goals may guide how you want to improve the product.

- **Be open about the tradeoffs:** Some candidates try to pass off their product as *the* best idea, as though their interviewer won't catch potential issues. Not so. Your interviewer will know the issues your idea has. If she doesn't, then you'll actually impress her by pointing out issues that she wasn't aware of.

When in doubt, act like you would imagine a good PM to act. A good PM will be realistic about the limitations. A good PM would gather research, but would also be able to move forward without it. A good PM will think about the business aspects. A good PM will be open about the potential issues with her designs.

Act like a good PM.

Sample Questions

1. How would you design a bookcase for children?

2. How would you design an oven for people in a wheelchair?

3. Google Maps is launching a version for schools. How would you design this?

4. What is your favorite business tool? Why?

5. How would you design a neighborhood park?

6. What would you change about a supermarket to make it better for college students?

7. What's your favorite picture storage website? What would you change about it?

8. Design a portal or interactive landing page to replace Google.com.

9. How would you design a social / career networking website for entrepreneurs?

10. Pick a target user who you don't feel is well served by Amazon.com. How would you redesign Amazon.com to appeal to that user type?

Case Questions
Chapter 15

It's often been said product managers are the "CEO of the product," but, unfortunately, the popular wisdom here is somewhat inaccurate. You are not the CEO. You will not do acquisitions or mergers. You will not deal with shareholders. You will often not even deal with finances.

However, you are responsible for delivering the best product to customers. Note this sentence covers three very important terms: "delivering," "product," and "customers." Most PM questions revolve around these three terms.

The Case Question: Consultants vs. PMs

If you have prepped for management consultant interviews, you should be aware that consulting case questions are just similar enough to PM case questions to potentially lead you astray.

Management consultants at firms such as McKinsey, Bain, and BCG have a different set of responsibilities than a PM. Consultants generally look at a firm broadly, spend weeks or months gathering data, and then tackle big-picture problems such as, "Should our client acquire _____?" They look at things like compatible company cultures, dependence on suppliers, and tax structures. They are expected to be data driven and methodical: ask questions, gather data, diagnose the issue, and propose a solution (which the consultants are rarely there to implement).

Product managers focus on the product and its customers. They make decisions about feature sets and market entry, and then follow through on implementing, delivering, and maintaining the product.

Even when interview questions overlap ("Should Google launch a TV service?"), appropriate interview behavior differs. Consultants will ask for (and be given) data to solve a problem while PMs will be expected to rely more on their instincts about an issue. For example, a PM interviewer is unlikely to whip out PowerPoint slides upon request. This is not to say that PMs should never be data driven or that consultants shouldn't use their instincts, but the focus is typically flipped.

Consultants: you have been warned.

What Interviewers Look For

One PM candidate I prepped explained that, in a recent interview with a top startup, he was asked what he would do if he were CEO of the company. "How can she even ask me that?" Michael complained. "I told her it was an unfair question since I obviously don't have as much knowledge as she does about the company." Suffice to say, that didn't go over so well.

For all the Michaels out there: Of course you don't have the knowledge to truly solve the problem. Of course you could solve it better if you did. But that's not the point; in fact, it's not even an issue.

An interviewer is not evaluating you on the correctness of your eventual conclusions but rather on the process you take to get there. For *that* goal, you don't need all the data and background information that you would have in the real world.

Interviewers are looking for candidates who will do the following:

- **Structure a problem:** Even seemingly open-ended questions can, and should be, broken down into components. Your market can be segmented and your strategy can be divided. Find a way to tackle a problem in a structured way.

- **Show strong instincts:** We rarely have all the data we'd like, and we don't know what the future will hold. A good PM should be able to make good business decisions, even in the absence of exhaustive data.

- **Drive, Not Ride:** You might not be the CEO of the product, but you are a leader. Show this by driving the interview forward. Be relatively exhaustive in your response to a question – discuss the benefits and tradeoffs, the short-term and long-term benefits, etc. – and back up your answers with reasons. Don't go overboard, though. If you find your interviewer is asking many follow-up questions and you're only giving short responses each time, you might be riding and not driving. On the other hand, if you're concerned you're going into too much detail, ask your interviewer if he'd like you to expand.

You can and should ask questions, but there's a limit. If your interviewer flips a question back to you (e.g., "Well, what do you think?"), that's a tipoff that you went too far with your questions. No big deal. Just get back in the driver's seat: think about the problem, make the best decision you can, and explain your reasoning.

Useful Frameworks

An MBA is not required for most product management jobs, and thus most business frameworks are not necessary to tackle problems. After all, many of your interviewers wouldn't know the frameworks either.

However, understanding them can be useful anyway. The frameworks can suggest a structure for a problem or point you to an area you might otherwise have missed. Just don't try to whip out the frameworks and force fit them to a problem. *Structure* matters, but regurgitating frameworks does not.

Don't worry about memorizing these frameworks; you won't be expected to know them. Just read them over to get a sense for the types of structure that can be used to answer a question. A framework is just a structure for taking a big, complicated problem and breaking it down into smaller pieces. Whenever you hear a Case Question, you should think about how you can break it down into smaller questions. These frameworks give examples of ways to do that.

Customer Purchase Decision Making Process

There are many frameworks to model the decision-making process, but two of the most common are AIDA and REAN.

AIDA models customer decisions as Attention (or Awareness) -> Interest -> Desire -> Action.

- **Attention:** You need to get the customer's attention somehow. A snappy email heading, perhaps? A snazzy ad? Or maybe a mention from a trusted friend or website?

- **Interest:** Now that you have the customer's attention, you need to get them interested in your offering. What are the advantages or benefits of your product?

- **Desire:** With the customer's interest piqued, you need to convince the customer that they want your product.

- **Action:** Finally, with the customer desiring your product, they take action to purchase the product.

REAN expands this to add on post-purchase behavior.

- **Reach:** The customer is aware of your product.

- **Engage:** The customer is engaged and considering your product.

- **Activate:** The customer takes action to purchase the product.

- **Nurture:** The customer has purchased the product, and it's now your responsibility to nurture this relationship.

Don't worry about which framework to use. Again, the specific framework isn't relevant. The point is to understand purchase behavior as a cycle wherein the customer must first be aware of your product, then they have to evaluate it against their needs and competitors' offerings, and then they finally purchase it. After the purchase decision, the customer relationship continues and your company should likely work to foster a strong relationship.

This framework might be useful in discussing how you would market a new product. You might discuss that getting the customer's attention will be fairly easy, but the "action" part (getting the user to actually switch from your competitor to you) will be more difficult.

Marketing Mix (4 Ps)

The "Marketing Mix" (also called the 4 Ps) is a way to understand the different aspects of a product's approach to marketing.

- **Product:** This is, of course, the actual item being offered. It should cater to a customer's wants or needs.

- **Price:** The price will determine how many and what type of customers purchase the product. Pricing can be more complex for online products and services as compared to physical products. For example, an online storage service could have a one-month free trial, followed by discounted pricing for non-profits, with additional "a la carte" purchases for an automatic backup utility.

- **Promotion:** Promotion encompasses all forms of advertising, PR, word of mouth, and sales staff. For example, promotion for a kids' product could include freebies given out to influential bloggers.

- **Place:** A physical product's distribution ("place") can include things such as online sales through Amazon, opening their own stores like Apple, distribution in retail stores, and sales through their own website. Greater distribution is not always better; many companies prefer to control the sales experience by limiting the sales channels. For online products, "place" might just be a single website, or it might include bundling the product with another company's offerings.

For online products, promotion can become very complex. A lot of products are competing for the customer's attention, and advertising is often insufficient to drive sales.

This framework could be useful to discuss the different elements of a marketing plan for a new or existing product.

SWOT Analysis

SWOT analysis is a structure to analyze companies and products.

- **Strengths:** Strengths are the *internal* factors that benefit a product. This can include anything about the costs, product features, company culture, reputation, infrastructure, or other aspects. For example, in considering launching the Kindle, one of Amazon's strengths would be that it is already the place where customers buy books online.

- **Weaknesses:** Weaknesses are *internal* factors that introduce challenges for a product. For example, if a web company were to consider creating a physical device, a weakness might be that it doesn't have experience with distribution.

- **Opportunities:** Opportunities have an *external* focus and relate to factors such as market growth, technology changes, competition, and legal regulations. For example, the growing cost of healthcare creates an opportunity for a product that gives better insight into one's health.

- **Threats:** Threats are the *external* challenges a product faces. For example, the unpredictability in energy usage regulation poses a threat for many clean energy products.

The following matrix represents the SWOT structure:

	GOOD	BAD
INTERNAL	strengths	weaknesses
EXTERNAL	opportunities	threats

This framework can help decide not only whether a company should pursue an opportunity but also what strategies would further that pursuit.

The Five Cs (Situational Analysis)

The Five Cs provide an overview of the environment for a product or decision.

- **Company:** This encompasses all aspects of a company, including its products, culture, strategy, brand reputation, strengths, weaknesses, and infrastructure.

- **Competitors:** Competitors include direct competitors, potential competitors, and substitute products. For each of these, a discussion could encompass market share, tradeoffs, positioning, mission, and potential future decisions.

- **Customers:** This includes aspects such as demographics, purchase behavior, market size, distribution channels, and customer needs and wants.

- **Collaborators:** Collaborators include suppliers, distributors, and partnerships. A discussion here might include what makes particular collaborators valuable and how they enable success.

- **Climate:** Climate includes aspects such as regulations, technology changes, economic environment, and cultural trends. A hostile climate can kill a business decision, while a positive one can greatly facilitate success.

This framework can guide discussions on whether a company should launch a product and what the company's strategy should be.

Porter's 5 Forces

Porter's 5 Forces is a framework for industry analysis. This industry analysis can be useful for understanding a company's decision.

- **Rivalry Among Existing Competitors:** More competitors generally leads to more heated competition, as does more direct competition. If many companies make the same product and they are not strongly differentiated, this will generally drive down prices for everyone. Many things can influence rivalry, such as market growth (growing markets enable competitors to expand without fighting with each other for market share) and high costs to exit the market (companies are reluctant to leave).

- **Buyer Power:** If a company or industry has relatively few buyers (for example, only the government and big banks), or some buyers have a very disproportionate share of revenue, these buyers will wield considerable power. This power allows them to affect prices, feature sets, delivery timelines, and other aspects.

- **Supplier Power:** Like buyers, suppliers gain influence over a company if the company is heavily dependent on them. This commonly happens if a company has a component that it exclusively (or almost exclusively) purchases from a single source.

- **Threat of Substitutes:** Competition exists not just from direct competitors, but also from substitute products. For example, even if Amazon were the only seller of electronic books (and therefore there was no direct competition), the prices of e-books would still be influenced by "competition" from physical books.

- **Threat of New Entrants:** With few barriers to entry in an industry, companies are constantly vulnerable to competition. If they price their goods too high, another company will enter the market and capture market share. Barriers to entry can include things such as proprietary technology, massive economies of scale, strong brands, or anything that's very difficult to do.

Consider, for example, the PC market. Buyers have considerable power, as many sales come from just a few retailers. Suppliers also have considerable power since there are limited manufacturers of certain components and high switching costs in changing manufacturers. On the positive side, there is some differentiation between competitors and limited substitutes. The market has moderate barriers to entry (branding, etc.). There are worse markets to be in, but there are also many better markets.

This framework could be useful in discussing whether or not a company should enter a specific market. What's the industry like? If it's hyper-competitive, we might choose to avoid it.

Make Your Own Framework

If you think a lot of these frameworks sound sort of the same, you're not alone. They do overlap. Here's a way to see the differences:

- **Customer Purchase Decision Making Process** helps us understand the buying process and gives us an "entry point" for boosting sales.

- **Marketing Mix (4 Ps)** describes the different aspects of a company's marketing plan.

- **SWOT Analysis** offers a framework for analyzing a strategic decision.

- **Five Cs (Situational Analysis)** gives an overview of the environment around a product or company.

- **Porter's 5 Forces** describes what an industry as a whole looks like.

Use these frameworks as a starting point to analyze a problem and to signal key aspects you might otherwise miss, such as the value of nurturing an ongoing customer relationship.

In an actual interview, you should use the best "framework," or structure, to solve the problem. If that's one of these frameworks, great! But more often than not, you'll need to create your own. You might slightly tweak one of these, you might blend two or three together, or you might come up with something totally fresh.

Product Metrics

Discussions about product and company metrics can come up in many ways. They could be introduced in a question about diagnosing a hypothetical issue, or they could be used to explain why a particular decision was made.

In your discussions, be sure to consider which metrics are most important for a product. Any product will excel in some metrics and struggle at others.

Types of Metrics

Depending on your goals, you can break down the types of metrics in a variety of ways. We've divided them into user acquisition, activity, conversion & retention, and money.

Alternatively, you can divide them based on the customer lifecycle, as Dave McClure does in his "Startup Metrics for Pirates" presentation[1]: Acquisition, Activation, Retention, Referral, and Revenue.

User Acquisition

- How many users do we have?

- How (and why) has the user base grown overtime?

- How many active users are there? How do we define what an active user is?

- Where are users coming from? Are they referring their friends?

---

1   "Startup Metrics for Pirates." Dave McClure. 8 August, 2007. http://www.slideshare.net/dmc500hats/startup-metrics-for-pirates-long-version

- Which channels are the most effective in getting users?

Activity

- How many users are using feature X?

- What percent have completed a particular workflow?

- What are people saying about the product? Do they love it? Can you measure that?

Conversion & Retention

- What is the conversion rate (free to paid, visiting to signing up, etc.)?

- What is the churn rate?

Money

- What is the customer acquisition cost?

- How much does supporting a customer cost?

- How much money does each user bring in (average revenue per user)?

- What is the lifetime value of a customer?

- What is our revenue growth rate?

Measuring

Generally speaking, you should measure the change in a metric rather than the total volume. Measurements since beginning of time are neat, but not actionable. They're the sorts of things a company might show off to investors or to the press. They won't help the company make a decision.

Many of these metrics can and should be broken down by "cohorts." These cohorts could be based on gender, location, date registered, or a variety of other factors.

To actually gather data, you have many options:

- **Usability Testing:** This generally won't give you true metrics, but it can help you understand the why. Why are customers leaving your site at a certain point?

- **Customer Feedback:** Feedback can come in from social networks, customer support pages, or surveys. Like usability testing, this will generally be more powerful in offer context to understand metrics than in gathering numbers.

- **Traffic Analysis:** Tools such as Google Analytics can help companies understand how users are interacting with the website.

- **Internal Logs:** Logging information directly can help a company understand user behavior at a deeper level than simple traffic analysis.

- **A/B Testing:** A/B Testing can help a company understand the impact of a particular change by comparing the behavior of users who have a feature to those who don't. While it is an incredibly useful tool, it can also be misapplied. For example, rolling out a new chat feature to only a small percentage of users might give you misleading data about usage patterns. After all, I can't use chat if none of my friends are.

Interview Questions

While not all interview questions fit cleanly into one of the categories below, many do – or at least overlap several categories. Below are the major types of case questions and how to approach them. These types include questions on strategy, marketing, launching, brainstorming, pricing, and problem solving.

For each type of question, a structured approach is vital. Find a way to break down a problem into components and then tackle each part.

Strategy Questions

Strategy questions include asking what a company's strategy is or how to design a strategy. It's important to think about a product's strategy at two levels:

- **Micro:** What is the business model for the product? What steps is the company pursuing to succeed on that model? Will customers want it?

- **Macro:** How does this product fit into the greater vision of the company? Will it open up new opportunities? Does it secure an existing market?

Many products will be strategic on both levels. For example, Amazon Kindle makes money on its own, but it *also* helps Amazon further secure its position as the go-to place to buy content.

In discussing the micro- and macro- strategy, frameworks such as SWOT and the Five Forces might be particularly useful. Think about questions such as:

- What is the company's mission?

- What are the company's goals?

- What are the company's strengths?

- What are the company's weaknesses?

- How is the company leveraging its strengths or minimizing its weaknesses?

Thinking about these questions at the micro and macro level and then applying their actions to it could help you describe a company's strategy.

Potential strategies for decisions could include:

- **Diversifying revenue sources.** A company might want to diversify revenue sources so drastic changes in the market or the emergence of a new competitor doesn't tank the company.

- **Building Barriers to Entry:** Building barriers to entry helps keep new competitors out and, therefore, protect a company's revenue. For example, Facebook's ownership of the social network makes it difficult (though not impossible) to successfully compete in the event management tools space.

- **Being the "One-Stop Shop for _____":** A company might want to expand their product suite around a particular area as a barrier to entry for competitors. For example, due to its breadth of products, Amazon is seen as the place you go to for purchasing anything. Companies that threaten this in some way by leading in a particular product area are acquisition targets for Amazon.

- **Being the Low-Cost Leader:** Being the cheapest will cut your profit margins, but it will also make the industry less attractive to current and potential competitors. Note that there's a difference between actively trying to ensure your prices are the cheapest and being forced to cut your price to compete. The former is a pro-active strategic decision; the latter is a consequence.

- **Reducing Reliance on a Key Buyer or Supplier:** Reliance on a single buyer can force you to meet their demands, and reliance on a supplier could cause unexpected delays or drops in quality.

- **Testing a New Market:** In some cases, a company will enter a new market with a niche product as a way of testing it out, building its brand, and learning more about the territory. The "real" product comes later.

This list is just to give you a taste of what strategies could be underlying a decision. Many, many more strategies are out there. It might be useful during your interview preparation to keep a list of key strategies you see.

Example

Facebook bought Instagram for about $1 billion, even though Instagram was making no money. Why do you think Facebook did this?

Answering this question does not require a strong knowledge of Instagram. That's not what the interviewer cares about. A strong answer for someone who doesn't know much about Instagram might look like the following:

> Hmm, interesting question. To be honest, I haven't really used Instagram much, but I'm vaguely familiar with it. And I'm of course familiar with Facebook. Let me break this down by discussing the key components of the acquisition and then how those aligned with or threatened Facebook's mission.
>
> Facebook's mission is to connect people and help them share their lives. This acquisition involved acquiring three things: the company (employees), the product, and the users. Let's think about these with respect to Facebook.
>
> While I'm sure Instagram had some very talented employees, the company was still pretty small at that time. I can't imagine that was a strong driver of the $1 billion acquisition.
>
> The product is a bit more interesting. Instagram created a beautiful photo-sharing product, and this was probably pretty scary to Facebook. Photo sharing is really vital to a company whose mission is helping people share their lives; it was a big draw of Facebook and a key strength. But another company was truly excelling there and basically beating Facebook (at least in certain aspects) at its own game. What happens when people start increasingly using Instagram instead of Facebook for photo sharing? It's a big threat.
>
> The users are the other big part of the acquisition. As I recall, Instagram had a ton of users as of the acquisition and had essentially built its own social network. This is of course what Facebook is all about and, again, Instagram had succeeded there. What happens as their social network grows? Will other companies (particularly creative sites like Etsy) start integrating with Instagram instead of Facebook? The size of Facebook's user base is the big barrier to entry for competitors, and every social network loosens Facebook's grip a bit.
>
> Ultimately, it seems like what Facebook might have seen is this hot,

young startup which had suddenly started excelling in two areas of strength for Facebook: photos and community. Facebook probably felt they couldn't afford to risk those areas or even allow another major player there. Plus, there also may have been a fear of what Instagram might do next – or what might happen if, say, Google got their hands on Instagram instead.

This candidate might have missed some aspects of the acquisition (such as an opportunity for Facebook to excel in mobile). That's an understandable oversight for someone unfamiliar with Instagram or Facebook's mobile strategy.

Note how the candidate has structured her answer. She's broken down the acquisition into its components, and then blended in some aspects of a SWOT analysis: strengths, weaknesses, and threats. This is one way the earlier frameworks might be applied.

She then wrapped up her response with a clear, succinct conclusion.

Sample Questions

1. Describe the strategy behind Google entering the cell phone market.

2. Amazon has a number of independent websites that, in some ways, duplicate the functionality of Amazon: Zappos (shoes and clothing); Diapers.com (baby and kid needs); YoYo.com (kids' toys); Look.com (kids' clothing); Soap.com (toiletries); Casa.com (house products). What do you think their strategy is in maintaining so many different websites with such similar functionality? Do you agree or disagree with it?

3. If you were Amazon, would you launch service in India? Why or why not? How would you do this?

4. Amazon has ads on its website for products you can buy on other websites. What do you think is the strategy there? Is it a good idea?

5. Do you think Apple should sell non-Apple products in its stores? Why or why not?

6. If you were responsible for Microsoft phones, what would you do?

7. Which Google products or services don't make sense to you? Why?

8. If you were CEO of Yahoo!, what would your strategy be?

9. Imagine you're starting up a new social networking service. What would your strategy be?

10. Imagine you were considering launching two services which have similar revenues and costs. How would you decide which one to pursue?

Marketing Questions

Marketing questions are about product positioning, customers, and handling competition. The Marketing Mix (4 Ps) is particularly useful here in offering an idea of the elements you might discuss. Aspects of SWOT, the 5 Cs, and the Customer Purchase Decision Making Process also come in.

In tackling these questions, you might try an approach like the following:

1. **Understand the Company:** What are the company's goals? What is its mission? What are its strengths and weaknesses? What threatens it?

2. **Understand the Competition:** Break down the competition into segments. Who competes and in what market? How do they position themselves? How entrenched are they? What are their strengths and weaknesses?

3. **Understand the Customers:** Who are your customers? What desire or need is this product fulfilling? What is the purchase/usage behavior like (one time only, repeated purchases, etc.)?

4. **Understand the Landscape:** Are there legal issues? If you're focusing on a particular region, what makes this region unique (e.g., frequent power outages)? What external forces are shaping your market?

5. **Market Your Product:** Using the information you discussed in the earlier steps, decide on aspects such as the product and its positioning and how you will promote it (think about the customer decision making process here). Depending on your interviewer and the question, you might also want to discuss pricing and distribution.

Note how a good chunk of the response covers just background information. That's to be expected; the marketing plan should follow from that discussion.

Example Question

How would you market Windows Phone to developers (to encourage them to adopt it)?

> Well, let me first think a bit about what the market looks like right now, including the company, competition, customers, and general landscape.

» *As I understand it, Microsoft's goal with respect to developers is to increase the number of high quality apps. Windows Phone has fairly low market share right now, which is the main barrier to this happening: It's just not interesting as a development platform. On the other hand, its strength is that Windows is dominant in the desktop market. This means ease of development and a strong brand name, at least in some senses. This is something for Microsoft to leverage.*

» *The two biggest competitors are iPhone and Android. The iPhone is excelling in the higher-end market, which draws developers to the platform. The perception is that iPhone users download and spend more money on apps. The Android phone is interesting to developers due to its market share, but it can also be frustrating because of the large number of devices to support.*

» *The customers, in this case, are the developers – but obviously their decision will be somewhat based on what users want as well. There are a lot of types of developers: hobbyists, enterprise developers, startup developers, etc. Developers want a fast and easy language to build a product with and the ability to push out their product to as many users as possible.*

» *The last thing to discuss, before going into a potential marketing plan, is the environment. We want to think about any trends that might open up holes in the market. One trend is the relative gap in the enterprise market, left by Blackberry's decline. The iPhone and the Android have filled in that gap a bit, but there may still be some room there (particularly if you can get the backing of IT/system administrators). Another relevant part of the environment is the diversity of phones from country to country. I don't know the statistics on this, but I suspect there are some markets where iPhone and Android haven't really dominated yet.*

Now for the marketing plan: I think the key things we want to focus on are (1) making development on our platform as frictionless as possible; (2) ensuring the top apps on iPhone and Android also have a Windows Phone version; and (3) finding an opening in the market. Our positioning to developers is that we'll make it easy to develop, maintain, and sell your app.

» *As far as making development easy: We already have a leg up here, in that many developers already know Windows development. We need to create an excellent set of tutorials and sample code to teach Windows Phone development. Let's try, if possible, to offer amazing APIs that developers can stick in their app to build key functionality – like plug-and-play coding.*

» *Additionally, because developers might struggle to find help on the web, we need to offer that help ourselves. We need to foster a community around Windows Phone development and offer free development support.*

» *We need to tackle where iPhone and Android are weaker in a development environment. I haven't done much mobile development myself, but I've heard iPhone developers complain about the insufficient data on tracking where downloads came from. Let's excel in this area.*

» *Ensuring that the top apps from both Android and iPhone are on Windows will take very proactive outreach. We need to track what apps are popular on those platforms. If they don't have a Windows phone, give them one. We can possibly even help them with testing and support in order to bring the quality up to par, particularly if it's just a single developer and not a full company or startup.*

» *The biggest market opening is probably in the enterprise space, so this might be where we want to focus. We can push for enterprise developers and IT folk to specialize in developing and supporting Windows Phones; many of them already do Windows development, so this isn't a big stretch. If we offer better tools to support their employees, this might encourage them to pick Windows over iPhone and Android for company-provided phones.*

This plan is designed to communicate to developers that we have a full ecosystem of tools: tutorials, APIs, technical support, free devices, community, analytics, administration tools, etc. This will differentiate us in the eyes of developers; we're putting them first to make us the fastest and easiest phone to develop for and support.

This isn't necessarily the best or only acceptable marketing plan. There are many marketing plans that might be acceptable here—even ones that lead to an entirely different and contrary conclusion.

After all, you aren't provided with data, so you have to rely on what you know—or think—to be true. If you're misinformed about something, you could wind up making a recommendation that your interviewer knows would utterly flop. A good interviewer might mark you off a little for the misinformation (since industry knowledge can be important), but should evaluate the rest of your response taking your understanding to be "fact."

Remember the goal of these questions is showing you can apply structure to an open-ended problem, make good business decisions, and think wisely about a product's marketing. You can do all of these things even with "bad" information.

Sample Questions

1. What is a product you think is marketed well? How would you improve it?

2. How would you market Gmail in China?

3. How would you market the Android tablet?

4. What do you think of Google's marketing for its social products? How would you improve it?

5. How would you market the next version of Windows?

6. How would you choose a market for expansion for Amazon Web Services?

7. Discuss how you think Whole Foods is currently marketed. What works well? What could they do better?

8. How would you market a mobile app to track weight, calorie consumption, and exercise?

9. How would you market Google Docs to schools?

10. Describe how you would market a magazine subscription service for Kindle.

Launching Questions

Launching products is one of the most important duties of a PM, so it makes sense that interviewers would ask how you would do this.

To tackle these questions, the following approach might work well:

1. **The Product:** Discuss the vision of the product, along with its strengths, weaknesses, and risks. What does the product ultimately hope to achieve? What is its target market? What are the things that would worry you? Who is it competing against and how is it positioned against those?

2. **Launch Goals:** Determine the goals of the launch. Do you want as many users as possible? Do you want to ensure profitability upfront? Do you want to validate that the product meets the market's needs? Do you want to ensure a positive reaction (even at the expense of slower growth)?

3. **Launch Design:** Now that you know the goals, determine how you will achieve those. What market would be a good test bed? Will you control growth through an invitation system? Will you roll out just a limited version of the product so you can launch earlier? Will you try to make the biggest

possible splash and get as many users as possible?

4. **Launch Implementation (Pre + During + Post):** With the basic design set, you now need to actually describe how you would implement this design. Break down your approach into three phases: pre-launch tasks, launch tasks, and post-launch tasks. Launching is not just a day-of process.

Your launch implementation discussion should cover the following aspects, possibly for each phase of launch:

- **Target Market:** Will you initially launch for just a particular city or a particular school? What is your initial target market?

- **User Types / Components:** If there are multiple types of users (e.g., drivers and riders), we may launch to these users separately: at different times and in different ways.

- **MVP or Full Product:** Will you launch an early minimum-viable product (MVP), or will you wait until the product is fairly full featured?

- **Distribution:** What stores and sales channels will carry your product? What platforms will you launch on?

- **Rollout:** When will it launch to different types of users? Are you going to start with a private beta? Will invitations be required?

- **Buzz:** How will you build buzz around your product (or will you)? Who will you reach out to? How will you get them on board? What do you want them to say about your product?

- **Partnerships:** Are there any partnerships that will help you to achieve your vision?

- **Risks:** What are the risks of the product, and how can you mitigate them? Are there legal challenges? Do you worry about a backlash from customers about privacy changes?

In thinking about these decisions, it might be useful to think about the Customer Purchase Decision Making Process. Which parts will be the most challenging? How will you overcome that?

Some of the possible launch activities are depicted in the below figure[2].

---

2 "Launching a Startup? Plan Your Marketing Around These 3 Phases." Dunford, April. 22 July 2013. From http://www.betakit.com/launching-a-startup-plan-your-marketing-around-these-3-phases/

RocketScope.com

Example Question

Imagine you're building a service to connect people with a recommendations/ booking service for local service providers (plumbers, etc.) – like an OpenTable for service providers. How would you launch this product?

A candidate could respond with an answer like the following.

> Okay, just to get this straight, what I'm imagining is a place where people could search for, say, a plumber or an electrician and see reviews. They could probably request an open time slot (if that's possible) or post a job and get bids.
>
> When I think about this product, it feeds a clear need. I don't know of other direct competitors to this.
>
> One of the challenges we'll face – and thus one of the things we'll need to design our launch around – is that it's a chicken and egg problem. We have two types of users: service providers and clients. Service providers won't want to join until there are enough clients, and clients won't want to join until there are enough service providers.
>
> The other challenge I suspect we'll face is being top of mind when a user needs a service provider. We're trying to change user behavior, and that's always hard.
>
> We want to design our launch to handle these two issues. In fact, given the challenges we're up against and given the odds that we won't get things quite right the first time, I think it's important we launch early

with a minimum viable product.

To tackle the first problem, we'll want to establish a strong user base of service providers first in a focused area (ideally, an affluent area with a lot of homeowners). We'll do a controlled launch where we can throttle client signup if it gets too high relative to the number of providers.

To tackle the second problem, we'll need to make ourselves more top of mind. The best way to do this is probably by focusing on services that are more regularly used: housecleaners, drivers, gardeners, pool cleaners, etc.

I think we'll need to launch, at the minimum, both a web client and mobile client for providers, and one or the other for clients. The providers need both since the mobile client allows them to take bookings on-the-go, and the web will help support providers without phones. For clients, I'd suggest that we start off with a web client. I do think the mobile client is really important, but it requires building for multiple operating systems. Given that we need to get something out ASAP, I think it's best to just do one for now.

In the months before launch, we'll start with getting providers signed up. They are less likely to get turned off than clients are by a mismatch in supply or demand for providers.

We'll pull providers from two places. First, we'll pull them from review sites such as Yelp and from community blogs and forums. Second, we'll reach out to real estate agents. They tend to be a go-to place for advice for new homeowners. We want them to know about our product, and we also want their list of recommended providers. An active sales force will be necessary to do this.

For the launch to clients, we'll do an invite-only, slow launch. We want to incentivize people to invite their friends though, so we may offer some sort of credit to people who invite their friends.

To build buzz, we'll also want to engage with the local community. We could have physical signage at popular cafes, reach out to local blogs, use Facebook groups about the community, etc. A lot of areas have their own newspapers, so we could try that too.

Post-launch, we want to analyze metrics on signups, conversion, etc. One of the things I expect we'll find is that many people sign up and then forget to use us when an issue comes up. We will need to track this carefully, as this will be a big challenge for us.

It's not just a numbers game, though: We really need to build a strong relationship with the service providers. We want them to be advocates for the site. If they can encourage existing clients to use it, then those clients may use it for other services, too. We also need their feedback to enable us to perfect the site.

So, to recap, we're doing the following: We'll build a mobile and web client for providers, and just a web client for users. We'll solicit providers through Yelp, community blogs and forums, and real estate agents, paying particular attention to recurring services such as gardeners and cleaners. We'll launch in a localized community – Silicon Valley is probably a good candidate – getting as many providers signed up as possible. The service will be gradually rolled out to clients with an invite-only registration. We'll maintain strong relationships with providers and clients and make it a point to excel in customer service. This will ensure we get the feedback we need to refine our product.

This candidate has recognized that this is really two launches in one: a consumer launch and a provider launch. He has then walked through how to launch in both of those markets and discussed key problems we might face. Finally, he wrapped up with an overview of his overall plan.

Sample Questions

1. Describe a product you think had a successful launch. Why?

2. Describe a product you think launched poorly, despite being a good product. What would you have done differently?

3. How would you launch the Google Self-Driving Car?

4. Suppose Yahoo were to build a phone. How would you launch this?

5. How would you launch an electronics store for Amazon.com?

6. Google is considering a version of Google Docs designed for large enterprises. How would you manage this launch?

7. You're working for a company that helps people create better online ads. How do you launch this product?

8. You have built a superior water bottle: it keeps drinks hot and cool, it's sturdy, and it never spills. How would you launch this product?

9. You have developed a car that takes a different type of fuel. This new fuel source is very cheap, but works as well as standard fuel. How would you

launch this car?

10. How would you launch a grocery delivery service in India?

Brainstorming

Brainstorming questions are about creativity. Interviewers want to see that you aren't stuck in a mindset of just doing the same old thing. They want to know you can come up with "big, hairy, audacious ideas."

For some people, this is difficult. They're scared of looking stupid, or they just get flustered and think too linearly.

If you're asked a brainstorming question, try to:

1. **Suspend disbelief.** It's okay to name a few stupid ideas, particularly because so-called "stupid" ideas are often the crazy ideas that get interviewers excited. Getting caught up in "Is this even good enough to mention?" is a good way to kill your creativity. Let your imagination run wild!

2. **Think about strengths and key assets.** If you find yourself stumped coming up with more ideas, try to forget about what the product is and just think about what its strengths and key features are. For example, consider a grocery cart. Its strengths are that it can hold a lot and that it is relatively compact (carts can be "stacked" inside each other). A key feature is that it is on wheels. What can you do with those?

3. **Think about one vs. many.** The use cases when you just have a single object might be different than when you have many. For example, one golf ball could be used to play a game. A pile of many golf balls could be used to weight down a tarp in the wind.

4. **Think about as-is vs. with modifications.** An object can be used as-is or it can be used with modifications. If you can do modifications to the object, the scope gets much broader.

Ideally, you want to go for "structured creativity." Structure shows organized thinking and helps your interviewer remember your answer better. However, don't let your structure get in the way of your communication.

After you're done brainstorming, your interviewer might ask you to go into one of your ideas in more detail.

Example Question

Name some innovative ideas for a vending machine.

There are many ways to respond to this question. Here's one of them.

Interesting. Let me think for a moment.

Okay, so first, we can use a vending machine to sell things other than the standard food and drinks. We can sell basically any small object. We could use a vending machine to sell beer at a bar, or we could use it to sell small electronics, cosmetics, etc.

It could be used outside of just the standard "sell small object" situation, though.

What a vending machine is good at is distributing exactly one copy of something – without breaking it – in a totally automated way. The automation is interesting because someone doesn't have to be present. Where else could this be used?

It could be used as an alternative to stores. Perhaps the owners of a small store could use a vending machine as a way to sell a limited amount of their merchandise while they're closed. They could even use it to speed up the sale of some express items during busy times of the day.

Or, if there were a way of modifying the vending machine so that people could insert items back into it, this could be used to rent things such as phone chargers. The beauty of that is if the person doesn't return it, you have their credit card on file.

This might even be useful for situations where merchandize is too valuable to leave just sitting out. A lot of stores have pricier items such as razor blades in locked shelves, and customers have to get a clerk to unlock it. If people could just swipe their credit card and pay instantly, it'd be better for everyone.

Okay, let me try to think even more abstractly now. A vending machine has the following attributes or components:

» *A big glass window showing the items.*

» *A refrigerator.*

» *The ability to locate a specific item on a shelf in an automated way.*

> » *A credit card processor.*

> » *A change machine.*

> » *Some way to validate that cash isn't counterfeit.*

> » *A way to prevent people from sticking their hand in and grabbing a bunch of items.*

It's also very heavy and automated.

We could use the glass window to display other types of items – small pieces of art, perhaps. We can use the refrigerator as a normal fridge.

We could also combine some parts. For example, if we had an old vending machine lying around, it could probably be modified to let people withdraw cash from their credit cards.

This ability to locate items at a specific row and column is interesting. Presumably it could also do this for multiple items at once. This sort of automation could be very useful in some industries, particularly when they want to reduce human error. Perhaps for pharmacists or doctors to distribute drugs and know they have the right one.

I think that's a pretty good list for now. Would you like me to keep going, or is there one particular idea you'd like me to dive into?

This candidate went with a "broadening" approach. The answer started with a pretty straightforward list of use cases for a vending machine, which used the technology in "expected" ways. Then, later, the answer broke down the machine by component and discussed other ways of using those components.

He potentially could have gone even broader. He could have discussed uses for a group of vending machines, or he could have analyzed modifications to the machine that could make it behave in different ways.

Sample Questions

1. Suppose you're a manufacturer of paper clips and have realized people no longer need paper clips. How else could you market your paper clips?

2. Name as many uses as you can for a wine glass.

3. Name some ways you could integrate Amazon products/services into a car.

4. Imagine you had access to an enormous database that transcribed all speech

everywhere in the world. As soon as someone says something, it is automatically transcribed and put into the database. It is also trivially fast to query anything. What products or services could you build with this database?

5. How could you integrate internet into a gym? What products or services could you create or improve?

6. If you were the CEO of Nike, what new product line would you come up with?

7. You are working at Apple and instructed to launch a product that is not technical. What ideas can you come up with?

8. You are suddenly given a very large (1000+) number of old desktop computers with old CRT monitors. What could you do with them?

9. Name as many uses as you can for chopsticks.

10. How would you improve the experience of shopping in a supermarket? Assume that even farfetched technology is within reach (provided that it's not actually impossible).

Pricing & Profitability

Pricing questions are aimed to get you to maximize profit, which is of course the difference between revenue and costs. In an ideal world, we'd have supply and demand charts that express how sales volume would change based on the pricing. We would then be able to pinpoint exactly when profit is maximized.

In the real world, we don't have any of this data. The goal – to maximize profit – is still the same, but the process is different. People generally use any or all of the following to price a product:

*   **Cost-Plus Pricing:** Examine the costs of your product and set your price a little higher than that. This is tricky because there are generally fixed costs and marginal costs, so it's difficult to assess the cost per unit. Additionally, many online services don't have direct costs, and the costs don't determine whether this is a *reasonable* price. However, the cost of your product does suggest a minimum price (presuming you want to make a profit) and indicates something about your competitor's prices.

*   **Value Pricing:** Some products have a clear and direct value to the customer. In those cases, you might be able to estimate how much money/time you are saving (or gaining for) the customer and price accordingly.

- **Competitive Pricing:** A great number of products are priced by just looking at the competitors' prices. This is partially rational (because your customers might otherwise select your competitors) and partially due to laziness (because people don't know how else to price a product). Pricing lower than your competitors is not necessarily a good thing; it can signal lower quality to customers, and it might start a price war. However, competitive pricing can still be a starting point from which you decide to price higher if your product is positioned as a premium product.

- **Experimental Pricing:** In some cases, it's possible for a company to experiment with different prices and then correlate price with sales volume. Proceed with caution here, though; inconsistent pricing can frustrate or anger customers.

A thorough company might use cost-based pricing, value pricing, and competitive pricing to triangulate on a good price, and then tweak it slightly with experiments.

With these general approaches in mind, there are a number of pricing models to consider:

- **Free, Ad-Supported:** Many startups try this approach, but few succeed. Advertising alone is rarely enough to support a company, unless there is something unique about your product which makes advertising particularly effective.

- **Freemium:** In a freemium model, a basic level of the product is free but a premium version is paid. This can be good for attracting customers. However, you have to keep a close eye on your costs for supporting these free users, as well as on your conversion rate.

- **Tiered:** A company might offer multiple levels of pricing, segmented by volume, customer type, or features. You don't want to go overboard, though; too many tiers can be overwhelming for customers.

- **À La Carte:** A company can price each feature or service separately, letting the customer choose exactly which "upgrades" they would like. This can often lead to customers paying more than they would have for a bundled suite of features. Some customers will like this flexibility but others will be overwhelmed by it. The support costs of dealing with so many different suites of features can also be challenging.

- **Subscriptions:** Many services offer subscriptions to their product or service, particularly in the case of web applications. Some products are simultaneously available for purchase and as a subscription. This enables products to capture customers who only need temporary usage and may not be willing to make the upfront investment in a full purchase.

- **Free Trial:** Short-term trials can be a good way to let customers experiment with a product before the purchase, as a way to "hook" them. Trials can be bounded by time, number of uses, or particular features (e.g., you can import but not export). You have to be careful to ensure a good enough experience that customers will enjoy the product but not so good that they don't desire to upgrade.

- **Razor Blade Model:** A company can sell one component (e.g., razors) at, near, or below costs with the expectation that an add-on component (e.g., razor blades) will bring in additional revenue. This can work very well if the customer can only buy these add-on components from you. If there are other competitors with compatible add-ons, then you run the risk of customers purchasing the product from you cheaply and then the add-ons from your competitor. No one wins (except for the customer).

A pricing model could use a combination of many of these attributes. For example, a company could offer subscriptions to its service, priced differently depending on the size of the customer's business, with additional upgrades purchased à la carte.

Online Advertising

As many tech companies (most notably Google and Facebook) have advertising-based business models, pricing questions around advertising come up particularly often.

Questions about advertising can seem complicated, but actually they usually break down into a handful of standard ratios such as click-through-rate (CTR), cost-per-click (CPC), and conversion rate. Once you start thinking about the problem in terms of those ratios, getting to an answer is usually pretty easy.

You will need to understand the basics of how online advertising works, though. Here's a quick overview.

Online advertising is generally priced one of two ways:

- **Pay-Per-Click:** An advertiser pays only when a person clicks on their ad. This means that if their ad is shown 1,000 times but only clicked twice, the advertiser would only pay for those two clicks.

- **Pay-Per-Impression:** An advertiser pays every time the ad is shown, regardless of whether or not the user interacts with the ad.

- **Pay-Per-Action:** An advertiser pays only when an action is performed (e.g., the customer clicks on the ad and purchases the product). This is also called a "conversion." This is a rarely used model, in part because of the difficulty of tracking conversions.

Google and Facebook offer both pay-per-click and pay-per-impression advertising. Additionally, Facebook offers pay-per-action advertising, with the allowable actions including "liking" a page, claiming an offer, and installing a mobile app.

At Google and Facebook, as well as other companies with advertising, the price of an ad is typically determined via an auction.

Cost-per-click advertising is the type more commonly discussed. The price you pay for each click is called the cost-per-click (CPC). In the auction, you bid based on the CPC you're willing to spend.

Another key metric is the percent of times your ad is clicked, called the click-through-rate (CTR). To calculate how much you'd pay to have your ad shown one thousand times, a metric called CPM, you can multiply: CPC * CTR * 1,000. For example, if the cost-per-click (CPC) were $1.50, and the click-through-rate (CTR) were 2%, then the CPM would be $30.

Generally, the most an advertiser would be willing to pay for an ad (that is, a click on an ad) is the expected profit from that click. To calculate the expected profit from a click, you want to know what percent of people who click on that advertisement will actually make a purchase (conversion rate). Then, multiply conversion rate * profit per conversion.

```
max for ad
    = expected profit per click
    = conversion rate * profit per conversion
```

If you were asked to calculate how much you would pay for advertising for a particular product, you might need to estimate the conversion rate and the profit per conversion. Conversion rates vary substantially by industry but are usually between 2% and 5% on Google search ads.

Example Question

How would you price a personalized notebook?

Consider the following response:

> Since it's personalized, I'll assume this is a premium notebook, probably leather (or faux-leather) bound.

First of all, let me think about who the market is. It feels like the market for this is probably consumers, but it's possible there's also some market for companies as a promotional item. Both would be good avenues to discuss.

There are a few different aspects of pricing to consider.

» *How much do nice moleskin-like notebooks cost? I've bought one or two, and they're fairly expensive—around $15 - $20.*

» *How much do other personalized products cost? We're competing in this market too, so we need to be aware of this. Personalized products can obviously be very expensive, but you rarely see personalized products for much less than $10. When I've looked for personalized gifts, there were a lot around $15 - $50.*

How much do our customers generally spend on gifts? A $20 gift is about the low-end of gift buying, at least for professionals (who are the most likely people to buy this product).

What sort of premium will people pay for personalization? We could survey customers to guess at this, but that's only somewhat accurate. A 50% - 100% premium seems about right.

How much does it cost us to make it? I have the least visibility into this. However, based on the prices of other products, I'm guessing that it takes at least the cost of the notebook itself, plus the cost of personalization. If a moleskin notebook is sold online for $20, then it's probably sold to the distributor for a 50% or so markup. The distributor also probably has profit margins of around 50%, so let's say that a moleskin notebook costs $5 to make. Personalization probably adds another few dollars on top of that.

It feels as though we're narrowing in on a pricing of about $35 for a nice, personalized notebook.

I suspect there's a market there for companies as a promotional item with the company logo engraved. We can offer volume discounts on large purchases of the notebook. This would incentivize large sales and offer an alternative to a lot of less classy promotional items.

Volume discounts wouldn't drop the cost of production of the notebooks (since those were already produced at a discount), but it probably does drop the price of engraving. Our costs are probably around $7, so that gives a lower bound on our price. For high volume sales, a price of $15 *seems right; that gives us 50% profit margins. We probably don't want to go much below this.*

The rest of the prices follow roughly from there:

» *1 – 10 notebooks: $35. I suspect the economies of scale don't kick in until after this point.*

» *10 – 30 notebooks: $30.*

» *30 – 100 notebooks: $20.*

» *100+ notebooks: $15.*

Of course, we'd want to tweak the prices based on data and actual sales. Since it's an actual physical product without ongoing customers to support, we could change the prices periodically without people getting too angry."

This candidate has used a variety of approaches to understand what a reasonable price is. She's evaluated how willing a person would be to pay a particular amount, based on the alternatives. In this case, alternatives include other notebooks, other personalized options, and other gifts in general. She has also validated this price against an estimation of the costs.

Sample Questions

1. Describe a product or service you think takes an interesting approach to pricing.

2. You are launching a new music service where you pay per time you play a song. After a certain number of plays, it automatically purchases the song. How would you price this service?

3. Imagine United Airlines has decided to launch a new "standby membership pass," which allows members to take unlimited free flights on standby (only boarding if there are empty seats). How would you price this membership?

4. A textbook publishing company has decided to sell subscriptions to electronic versions of any of its books. How would you determine the price of this service?

5. How would you price a file backup service targeted at enterprise customers?

6. Two stores are located just a mile away from each other. They sell the same product, but one sells the product for 25% more than the other. What factors

could explain why the prices are different?

7. How would you price a cell phone service for the elderly?

8. You are launching a photo printing service that automatically prints and mails someone's favorite pictures to them. How would you price this service?

9. Imagine you are launching a to-do list manager for smartphones. How would you determine the pricing of this application?

10. What analysis would you go through to determine if/when Amazon should change the price of Amazon Prime?

Problem Solving

Some questions are left wide open: There's a problem. What would you do?

Tackling these problems starts not with the solution but with the problem. You have to isolate the precise source of the problem, then diagnose the cause, then solve it.

In some cases, you might be ultimately led to an irreconcilable conflict between two paths. For example, suppose you have a question like, "You are launching a photo sharing site and have experimented with a new interface for posting pictures. It increases time spent on the website, but decreases the number of photos shared. What would you do?"

If you must make a choice between two paths, then let the company's goals be your guide. The goals will vary not only from company to company, but also within the same product at different times. For example, a new startup might value user acquisition the most at first, but then later prioritize revenue.

Depending on the interviewer's goals, she might conduct this as a back-and-forth exercise, where you ask questions to clarify and she provides answers. Or she might want to see your instincts in general and to see how you would solve the issue in the real world.

Breaking down the problem into components can help you isolate the causes. Some of the common problems and their causes are:

- **Falling profit** results from a decline in revenue or an increase in costs.

- **Falling revenue** results from a decrease in sales volume or a decrease in price. This could mean a *shift* in purchasing behavior across tiers of the product.

- **Falling sales volume** results from a decline in new customers coming in or lower retention from existing customers.

- **Declining new customers** results from a decline in traffic or in conversion rate. Either of these could come from repeat visitors versus new visitors on a website.

- **Increase in costs** can be caused by an increase in fixed costs or an increase in marginal costs. These, in turn, can be caused by suppliers increasing their prices, a distributor changing their profit structure, a spike in the number of returns, or a variety of other aspects.

- **Decline in traffic** can be the result of a decline in the number of new visitors, a decline in the number of returning visitors, or a decline in the engagement for either of those types of users.

- **Decline in new visitors** can be the result of a decline in search traffic, a decline in referral traffic, or a decline in direct traffic.

A question like, "How would you figure out why our profit has declined?" could be broken down from profit, to revenue, to sales volume, to new customers, to conversion rate. The key is to isolate the problem.

Once you've figured out which variable is actually changing, you then need to diagnose the problem. This could be almost anything, but here are some ideas to get you started:

- Has this happened in all regions?

- How many product lines do we have? Has this happened in all our products?

- Have competitors had similar issues, to the best of our knowledge?

- Have other related products experienced the same effect?

- Have we seen any seasonality?

- Have we made any changes to our product line?

- Have new competitors entered the market?

- If we separate our customers by new versus returning, what differences do we see?

- How is customer retention?

- What have customers been saying? Have we been getting more complaints recently?

- Do we notice any changes in referral traffic?

Once you've diagnosed the problem, you might be asked to resolve it. Clarify the goals before you go on to doing this.

Example Question

You are working for Amazon in their clothing category when you discover that the sale of jeans has steadily declined during the past three months. How would you figure out what happened?

You might approach this question as follows:

> *Interesting. Okay, so the sale volume has dropped. I'll assume this was a substantial drop. I need to understand more about what happened.*
>
> *The three month thing is particularly curious. It seems that something happened with jeans, with clothing, with Amazon, with online commerce, or just with time.*
>
> *I think it's fair to rule out a big issue with online commerce or Amazon. com. Presumably, if those had experienced major drops, we'd know about it and have bigger issues to worry about. Likewise, let's rule out a decline in clothing on Amazon since, presumably, we wouldn't be so concerned about jeans if it were a clothing-wide issue.*
>
> *The time issue is interesting. Comparing sales to three months ago isn't necessarily a good idea. Jeans are probably not sold at a steady rate year round. It'd be better to compare the sales to a year earlier to rule out seasonality. Let's assume, though, that this still reveals an issue.*
>
> *You've also said it's a steady decline. This means that it's probably not, for example, a single UI change that caused the issue. That would cause a sudden decline instead.*
>
> *At this point, I'd like to get an understanding of how the sales have been within the jeans category.*
>
> » *Have all jeans faced a decline? Or just particular brands?*
>
> » *Has Zappos experienced a similar decline with jeans? Since Amazon owns Zappos, we could likely get this information.*
>
> *Let's say it was all jeans on Amazon, but that Zappos did not experience*

an issue. This isolates the problem to something specific that we did.

Sales volume is a function of visitors and conversion rate. If sales dropped, then one or both of those must have changed.

We could break down both visitors and conversion rate by type of visitor: Amazon search traffic (people searching on Amazon.com), browse traffic (people browsing to get to the right category), external search traffic (search engine traffic, such as from Google), and direct traffic (people going directly to a particular product).

Our goal is to look for which of these experienced a decline either in the number of visitors or in conversion rate. If, for example, external search saw a drop in traffic, then we could investigate what changes we had made that might have affected our pagerank.

If we don't see a difference within these types of users, then we'll want to look elsewhere for something that might have impacted sales. For example, it might be that the price of jeans increased substantially. It's also possible (though hopefully unlikely) that there was an error in the searching/buying process that caused a decline.

This is the basic process I'd go through. Is there a particular aspect you'd like me to drill into further?

This candidate has paid close attention to the wording of the question. It's a *gradual* decline across a type of product. This should be a clue to what might be causing the problem.

The candidate has also struck a good balance between outlining a general approach to solve the problem and using appropriate instincts to guide the direction (e.g., it's unlikely that jeans suddenly stopped selling well).

Sample Questions

1. You notice the advertising revenue on your website has dropped considerably. How would you go about figuring out why that has happened?

2. You compare traffic from this month to last month and discover that this month's traffic is 10% lower. What would you do?

3. A magazine company comes to you for help. They understand publishing is a troubled industry. However, their sales have declined 10% while those of their closest competitor have declined just over 5%. How would you handle this issue?

4. A particular page on Facebook results in an error 10% of the time. What could cause this?

5. Your VP demands that you double revenue within four years. How would you go about creating a plan to do this?

6. You own a small ecommerce website that specializes in selling sporting goods. Last year you made almost $200k in profit. This year it was just $80k. What could have happened, and how would you figure out the cause?

7. You own a website with three tiers of pricing: free, standard, and premium. What would you do if you see that the sales of the premium product have fallen but those of the standard product have increased?

8. Two children are running lemonade stands, just a few blocks away from each other. What might cause one child to do much better than the other?

9. You are about to launch a major change to the user interface of your company's website. What sort of metrics would you want to monitor to notify you if there's a problem?

10. You notice that the Google AdWords revenue for a particular word has dropped in Spain for the last 30 days. This is an important one, so Google has asked you to figure out what has happened. How would you tackle this?

Coding Questions
Chapter 16

Thought you'd get away without having to code? Not so fast.

Many companies, including Google, Amazon, and Microsoft, sometimes ask their PM candidates coding and algorithm questions. These questions can range from straightforward coding questions to more complex algorithm questions (which may or may not be followed up with a request to code).

The good news is the expectations are generally lower than they would be for a developer. Many, but not all, interviewers will be satisfied with pseudocode.

Who Needs To Code

Generally, the more recently you've coded, the more likely you are to be asked coding questions. Companies will often expect recent computer science graduates and current developers to code during their interviews.

Never coded before? You're probably off the hook for the coding aspects, but you still might be expected to explain an approach to solve a problem. It's worth reading this section, just in case. Some of the terms might be beyond your knowledge, but you'll still benefit from understanding how to approach these questions.

What You Need To Know

If you're a recent graduate from a strong computer science program, you probably know most of what you need to know. Focus more on practicing interview questions than relearning actual knowledge, unless you discover major gaps.

Here's the quick list of what you should know.

> Note: we're jumping right into big O notation to explain runtimes. If you don't know big O or you are very vague on its meaning, you might want to jump ahead to read about it.

Data Structures

As its name suggests, a data structure is a structure for holding data. Depending on what you're optimizing for, there are many different approaches to holding or organizing your data.

In roughly descending order of importance for an interview, the common data structures are:

Arrays

An array is the most straightforward way to hold a set of objects. It stores items in a simple list of objects. Looking up an object is fast if you know the index, but slow otherwise. For example, it's fast to retrieve the 12th person in a list, but slow to find all people named "Alex" (since you have to look through all people).

In most languages, an array cannot "grow" in length after being created. You must specify the length of the array upfront and it cannot be changed after that.

Good Practice Problems:

16.1   Given a sorted array of positive integers with an empty spot (zero) at the end, insert an element in sorted order.
pg 303

16.2   Reverse the order of elements in an array (without creating a new array).
pg 305

Hashtables

A hashtable (sometimes called a "dictionary" or a "hashmap") allows you to map a "key" to a "value." This key is often a number or string, and the value can be any type of object.

This is a very useful data structure because it allows for very fast lookup. For the purposes of an interview, we generally assume that a hashtable is $O(1)$ (constant time, regardless of the amount of data) to insert and look up elements, even though this isn't 100 percent true. A poor implementation of a hashtable could have an $O(N)$ look-up time.

You might use it to map from a person's ID number to some object with other information about them.

Good Practice Problems

16.3 Given two lists (A and B) of unique strings, write a program to determine if A is a subset of B. That is, check if all the elements from A are contained in B.

pg 306

16.4 You are given a two-dimensional array of sales data where the first column is a product ID and the second column is the quantity. Write a function to take this list of data and return a new two-dimensional array with the total sales for each product ID.

Example:

Input:

211,4
262,3
211,5
216,6

Output:

211,9
262,3
216,6

pg 307

Trees and Graphs

A **graph** is a set of nodes which are connected through edges. Not all the nodes need to be connected—you could have two entirely separate subgraphs—and the edges can be either "directed" or "undirected." A directed edge can be thought of as a one-way street, with an undirected edge being like a two-way street. If the graph is directed, an edge from v to w is not an edge from w to v. Therefore, you might be able to "drive" from node n to node m, but not the other way around.

A **tree** is a type of graph in which any two nodes are connected through one, and only one, path. A tree will not have any cycles since there can only be one path between any two nodes.

A tree can come in many forms, but by far the most common is the **binary tree**. A binary tree is a tree where each node has only two child nodes. We call these nodes the left node and the right node. As with all trees, there cannot be any "cycles" on the tree (no paths from a node back to itself). Because of these restrictions, a binary tree can be represented in a strictly hierarchical fashion like this:

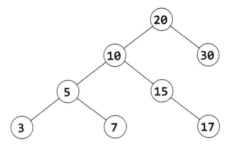

Commonly, we work with **binary search trees**. A binary search tree is a tree in which all nodes in the left subtree are less than the node's value, which is in turn less than the values of the all nodes in the right subtree. The above tree is a binary search tree.

If a binary search tree is balanced (and usually we deal with balanced binary search trees), inserting an element, as well as finding an element, is $O(\log n)$ where n is the number of nodes.

Good Practice Problems

16.5 Insert an element into a binary search tree (in order). You may assume that the binary search tree contains integers.

pg 308

16.6 Given a binary search tree which contains integers as values, calculate the sum of all the numbers.

pg 309

Linked Lists

Like a binary tree, a linked list is a data structure composed of nodes, where each node has a pointer to other nodes. In a singly-linked list, the node only has a pointer to its next node. In a doubly-linked list, the node has pointers to its previous node and its next node. It is generally considered highly problematic, and possibly a violation of the linked list structure, if the list has a cycle.

Inserting a node into the front of a linked list can be done in $O(1)$ time. However, if the list is sorted and you wish to insert the node in order, it will take O(N) time, where N is the number of nodes. This is because you must first find the right spot, and this requires searching through the whole list.

Finding a node in a linked list is O(N), whether or not the list is sorted.

Good Practice Problems

16.7 Insert a node into a sorted linked list (in order). (Don't forget about what happens when the new element is at the start or end!)

pg 310

16.8 "Sort" a linked list that contains just 0s and 1s. That is, modify the list such that all 0s come before all 1s.

pg 311

Stack

A stack is a data structure which defines a precise order for how elements must be inserted and removed. When an element is added, or "pushed," it is inserted on the top of the stack. When an element is removed, it is "popped" from the top of the stack.

A stack is said to be a LIFO (last-in-first-out) data structure, since the last (most recent) element added is the first one to be removed.

In this way, it acts like a stack of plates in real life. When you add a plate onto a stack of plates, you add it on the top. When you remove a plate, you always remove from the top.

Inserting and removing from a stack is O(1). Finding an element with a particular value is not usually done, as it would require removing all the elements, one by one. A stack is not a good data structure choice if this is something you think you will need to do.

Good Practice Problems

16.9 Write a function which takes a stack as input and returns a new stack which has the elements reversed.

pg 316

16.10 Write a function which removes all the even numbers from a stack. You should return the original stack, not a new one.

pg 316

Queue

A queue is essentially the opposite of a stack. Rather than removing the newest item with a LIFO (last-in-first-out) principle, it removes the oldest item. It is said to be "FIFO" (first-in-first-out), since the first item you add will be the first one you remove.

It acts like a queue (or line) in real life. When people are in a queue for movie tickets, the first person to get in line is the first person who will be served. This is of course how the data structure gets its name.

Inserting (or "enqueuing") and removing (or "dequeuing") from a queue is $O(1)$. As in a stack, finding an element would not ordinarily be implemented.

Good Practice Problems

16.11 Write a function to check if two queues are identical (same values in the same order). It's okay to modify/destroy the two queues.

pg 317

16.12 Write a function to remove the 13th element from a queue (but keep all the other elements in place and in the same order).

pg 318

Algorithms

If you have a computer science degree, you know we could easily fill hundreds of pages with advanced algorithms. We won't though, because these topics are rarely asked in interviews. Even developers are unlikely to be asked questions about, say, Dijkstra's Algorithm, since interviewers care much more about your ability to create a *new* algorithm than memorizing an existing one.

Still, there are a few fundamental algorithms that are considered "fair game" for developers, and even for PMs. These come up frequently enough that it's worth your time to remember them.

Sorting

The two most common *good* ways to sort an array are quick sort and merge sort. The others—bubble sort, insertion sort, radix sort, etc.—are less efficient in general or only work with specific assumptions.

- **Merge Sort** operates by sorting the left and right half of the array, and then merging the arrays. How does it sort the left and right halves? Through the merge sort algorithm (recursively). It takes the left half, divides that in half, sorts each part, and then merges those together. It then does the same on the right half. Merge Sort is O(n log(n)) in the average case and in the worst case.

- **Quick Sort** sorts data by choosing a random "pivot" element and rearranging the elements in the array based on whether they're less than or greater than the pivot. Next, it tackles the elements on the left side of the pivot (all of which are less than or equal to the pivot) and the right half of the pivot (all of which are greater than the pivot). It applies the same strategy to each side: pick a pivot, rearrange, and then pick a new pivot on each side. Quick Sort is O(n log(n)) in the average case, but O(n²) in the worst case. The worst case will happen if a bad "pivot" (a very low or very high element) is continuously picked.

Note that both algorithms have an approach of "divide in two parts and then re-apply algorithm."

The other sorting algorithms are the naive implementations that you might do when, say, you're trying to sort a stack of papers.

- **Insertion Sort** maintains a sorted sublist of elements (initially 0) at the beginning of the array. It then looks at the beginning of the unsorted sublist. If this element is bigger than the last element in the sorted sublist, it leaves it in place and just grows the sorted portion (since the element is already in the correct order). If it's smaller, then it moves it into place in the sorted sublist. The unsorted portion shrinks by one each time. The algorithm then repeats this step for each element in the unsorted portion, until the array is fully sorted. Insertion Sort takes O(N) time in the best case (if the array is already sorted), but O(N²) in the expected and worst case.

- **Bubble Sort** is a pretty straightforward algorithm. It iterates through the list repeatedly, swapping each pair of elements that are out of order. Once a full iteration happens without any swaps, the array is sorted. This takes O(N) in the best case (if the array is totally sorted) and O(N²) in the expected and worst case.

It's unlikely that you'll be asked to implement one of these algorithms by name, but it's still useful to understand how one might sort data. PMs are sometimes asked to sort a list of data. Those with a recent CS degree would likely be expected to implement one of the more optimal algorithms, while those without a CS degree may get away with the more naive approaches.

Good Practice Problems

16.13 Given two sorted arrays, write a function to merge them in sorted order into a new array.

pg 319

16.14 Implement insertion sort.

pg 320

Binary Search

Binary search is an algorithm for locating a value in a sorted list (typically an array). In binary search, we compare the value to the midpoint of our list. Since our list is sorted, we can then determine whether the value should be located on the left side or the right side of this comparison element. We then search the left or right side, repeating this operation: compare to midpoint of the sublist, and then search the left or right half of that sublist.

Because we're repeatedly dividing the data set in half, the algorithm takes O(log n) time in the average and worst case.

We often perform binary search in real life without realizing it. Imagine you had a stack of student exams sorted by first name. If you had a name like "Peter," would you search starting from the top of the stack? Probably not. You would hop about halfway through, and then compare. If you see "Mary," you know to keep going. You could then just search that second half of exams by continuously dividing the stack in half.

Binary search is a popular algorithm and therefore an important concept to understand. Many algorithms are based on binary search.

Good Practice Problems

16.15 Implement binary search. That is, given a sorted array of integers and a value, find the location of that value.

pg 319

16.16 You are given an integer array which *was* sorted, but then rotated. It contains all distinct elements. Find the minimum value. For example, the array might be 6, 8, 9, 11, 15, 20, 3, 4, 5. The minimum value would obviously be 3.

pg 322

Graph Search

There are two common algorithms for searching a graph: depth-first search and breadth-first search.

In depth-first search, we will completely search a node's first child before going on to the second child, third child, and so on. For example, imagine a node with two children, A and B. If we are searching for a value v, we completely search A (and the nodes connected to A) before we check out B. It's called "depth-first search" for this reason; we go deep before we go wide.

In breadth-first search, we go wide before deep. If we start from an initial node R, we first check R and all the nodes immediately connected to R (let's call these "children"). Then, we expand our search outwards, searching all the nodes connected to R's children. We repeat this process until we find the value or until we've completed searching this entire [sub-]graph.

In both algorithms, we need to be careful that we don't wind up going in circles forever. Therefore, if there are cycles in the graph—that is, if there is more than one path to get from one node to another—then we need to mark the nodes as "already visited" to ensure that we don't repeatedly search the same node. This will not be an issue for trees, as there are no cycles in a tree.

Note that a graph *can* have two completely separate parts that are not connected. If this is the case, we need to perform our search algorithm on each component to ensure that we find the item we're looking for.

Good Practice Problems

16.17 Using depth-first search, check if a tree contains a value.

pg 324

16.18 Write the pseudocode for breadth-first search on a binary tree. Try to be as detailed as possible.

pg 324

Concepts

Big O Notation

Big O notation is a way to express the efficiency of an algorithm. If you're going to be working with code, it is important that you understand big O. It is, quite literally, the language we use to express efficiency.

Big O will allow you understand the tradeoff of different features. For example, if you were working on a social networking website and you wanted to show how many friends two people have in common, you might suggest looking through each of my friends to see if the friend is in your list of friends. This probably takes $O(N^2)$ time, where N is the average number of friends a user has. That is, if you were to time how long this approach took as friends lists grew bigger and bigger, you would see that the graph of runtimes looks something like the $f(x) = x^2$ graph. This is going to be very costly. You'll need to come up with a better implementation.

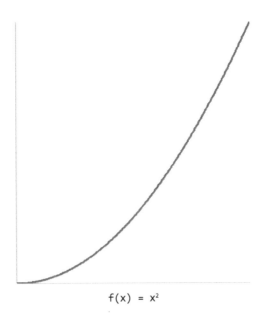

$$f(x) = x^2$$

Big O allows you to state this sort of information clearly and succinctly. It expresses how the execution time of a program scales with the input data. That is, as the input gets bigger, how much longer will the program take? Just a little bit longer? A lot longer? Will the time increase, say, exponentially with the size of input (yikes!)?

Suppose you have a function *foo* which does some processing on an array of size N. If *foo* takes $O(N)$ time, then, as the array grows (that is, as N increases), the number of seconds *foo* takes will also increase in some sort of linear fashion.

Carrier Pigeons vs. The Internet

This is a true story.

In 2009, a South African company named The Unlimited grew frustrated by their ISP's slow internet and made news by comically showing just how bad it was. They "raced" a carrier pigeon against their ISP. The pigeon had a USB stick affixed to its leg and was taught to fly to an office 50 miles away. Meanwhile, the company transferred this same data over the internet to this same office. The pigeon won—by a long shot.

What a joke this ISP was, right? A bird could transfer data faster than them. A bird!

Their internet may or may not have been slow, but this experiment doesn't say much. No matter how fast or slow your internet is, you can select an amount of data that will allow the internet or a pigeon to win.

Here's why:

How long does it take a pigeon to fly 50 miles with a 10 GB USB stick attached to its leg? Let's say it takes about 3 hours. Great.

Now, how long does it take to transfer 10 GB on the internet? Let's say you have pretty fast internet, and 10 GB only takes 30 minutes. Okay, then transfer 100 GB and you know it will take more than 3 hours.

How long does it take that same pigeon to "transfer" 100 GB? Still 3 hours. The pigeon's transfer speed doesn't depend on the amount of data. USB sticks are pretty light but can fit a ton of data. (This is a bit of an oversimplification, of course. With enough data, you would need many USB sticks and eventually many pigeons.)

So, just like that, the pigeon beat the internet!

The pigeon's transfer time is constant. The internet's transfer time is proportional to the amount of data: twice the data will take about twice as much time.

In big O time, we'd say that the pigeon takes $O(1)$ time. This means that the time it takes to transfer N gigabytes varies proportionally with 1. That is, it doesn't vary at all.

The internet's transfer speed is $O(N)$. This means that the amount of time it takes varies proportionally with N.

Big O offers an equation to describe how the time of a procedure changes relative to its input. It describes the trend. It does not define exactly how long it takes, as procedures with larger big O time could be faster on specific inputs.

Real-Life Big O

Many "operations" in real life are $O(N)$. Driving, for example, can be thought of as $O(N)$. As the distance *N* increases, driving time also increases in a linear fashion.

What might not be $O(N)$?

Imagine we invited a bunch of people (including you) to a dinner party. If I invited twice as many people to the party, you will have to shake twice as many hands. The time it will take *you* to shake everyone's hand can be expressed as $O(N)$. If I double the amount of guests, it will take you twice as long. This is a linear, or $O(N)$, increase.

Now, let's suppose everyone wants to shake hands, but for some strange reason only one pair of people can shake hands at a time. As N increases, how much longer will this meet and greet take? Well, your work will take O(N) time—but so will everyone else's. The time it takes increases *proportionally* with O(N^2), since there are roughly N^2 pairs.

Dropping Constants

If you are paying close attention, you might say, "But wait! There aren't N^2 pairs. People aren't shaking hands with themselves, and you're double counting every pair. There are really N(N-1)/2 pairs. So we should say O(N(N-1)/2)."

You're absolutely right. There *are* N(N-1)/2 pairs (which is .5*N^2 - .5N), but we still say that this is O(N^2).

Big O is very hand-wavey, wishy-washy. We're trying to express how the time changes in rough terms, not offer a precise calculation for the number of seconds something takes.

As a result, we drop constant factors, so O(2N) is the same as O(N). We also drop the addition or subtraction of constants, so O(N - 5) becomes O(N). Put together, these two factors mean that O(N^2 + N) should be written as O(N^2). Think about it: if O(N^2) and O(N^2 + N^2) are the same, then O(N^2 + N), which is between those two, should be treated as the same.

This is a very important thing to understand. You should never express an algorithm as "O(2N)." This is not a "more precise" or "better" answer than O(N); it's only a confusing one. A so-called "O(2N)" algorithm is O(N) and should be expressed as such.

Which of the below expressions are equivalent to O(N^3)?

```
O(3N³)
O(N(N² + 3))
O(N³ - 2)
O(N³ + N lg N)
O(N³ - N² + N)
O((N² + 3)(N+1))
```

All of them!

Drop your constants and just keep the most important term.

Multiple Variables

Back to the handshaking example. Suppose we invited men and women to our dinner party. All the men already know each other and all the women already know each other. Therefore, people will only shake hands with the opposite gender.

Assuming that we're still in bizarro land where only one pair can shake hands at a time, how would you express how long this takes?

Don't say $O(N^2)$. Suppose we have 100 men and 1 woman. Adding one man will add one handshake, but adding one woman will add 100 handshakes. The time it takes does not actually increase proportional to the number of people squared.

These are different "variables," and it matters which one we increase. The correct way to express this is with two variables. If there are M men and W women, then our meet and greet takes $O(M*W)$ time.

What if the women all knew each other, but the men knew no one at all? We would then say that the meet and greet is $O(M^2 + M*W)$. Note that we do not drop that extra M*W term; it's a different variable, and it matters.

Why This Matters (And Why It Doesn't)

Let's suppose that we have two functions which process some data. The function *foo* takes $O(N)$ time and the function *bar* takes $O(N^2)$ time. On a given data set (for example, a specific list of people), which one will be faster?

We don't know, actually.

The runtime of foo will increase proportionally to $O(N)$ and the runtime of bar will increase proportionally to $O(N^2)$. So, eventually, the $O(N^2)$ line should exceed the $O(N)$ time.

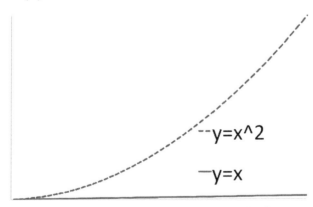

However, we can't make any determinations on a particular data set. The $O(N^2)$ could be faster on smaller data sets; it might not have yet exceeded the $O(N)$ line. Plus, even on very large data—after this "overtaking" occurs—there could be exceptions. Maybe, when N is divisible by 1000, the *bar* code will hit a special case and suddenly operate very quickly. We just don't know.

This doesn't make big O useless; we just have to be very careful about how we apply it.

Big O allows us to say things like, "In general, as our data set grows in size, this algorithm will be much, much faster than this other one." It also allows us to say, "You want to run this $O(N^2)$ algorithm, and N is the number of files on our network? Sorry, that's just not going to work." That matters—a lot.

Moreover, it gives us a language for expressing efficiency that isn't reliant on the system architecture or the technologies used. Without big O, we'd likely have to discuss efficiency in terms of seconds, which has little meaning when you are on a different system.

Logs and Big O

You might notice, as you're doing problems, that we (and others) describe problems as $O(\log(N))$ or $O(\lg(N))$, but we aren't particularly concerned about specifying whether we mean $\log_2(N)$ or $\log_{10}(N)$. That's because it doesn't matter. The difference between one log and another is just a constant factor: $\log_b(n)$ equals $\log_k(n) / \log_k(b)$. Since big O time doesn't care about constant factors, we don't need to care about what our log base is.

Big O Space and More

The concept of big O can be used for much more than runtime. In fact, very commonly it is used to describe how much memory an algorithm uses.

For example, suppose I have an algorithm that creates and initializes an NxN matrix:

```
1   int[][] a = new int[N][N]; /* NxN matrix
2   for i from 0 to N {
3      for j from 0 to N {
4         a[i][j] = i + j
5      }
6   }
```

This algorithm takes $O(N^2)$ time and $O(N^2)$ space.

> *Note: If you've taken an algorithms class, you might remember that, technically, big O refers to an upper bound. Anything that is O(N) could also be said to be O(N^2). To describe the exact runtime, we should be using big-theta.*
>
> *That is true, by the official mathematical definition of big O. However, out-*

CRACKING THE PM INTERVIEW | MCDOWELL & BAVARO

side of an algorithms class, this distinction has been forgotten about.

Sample Problems

Now, let's move on to some examples (in pseudocode). Can you find the runtime of each of these problems?

Example 1

Consider the following code to print the numbers from 0 to n.

```
1   for i from 0 to n {
2       print i
3   }
```

This is said to be O(n) time. That is, if we were to run this code for many different values of n, the runtime would increase at a rate proportional to n.

Example 2

What about this code?

```
1   sum = 0
2   for i from 0 to n {
3       sum = sum + i
4       for j from 0 to n {
5           sum = sum + j
6       }
7   }
```

This is O(N²) time. There are two for-loops, each running from 0 to n. How many times does line 5 get executed? O(N²). The time for this code to run will increase at a rate of O(N²).

Example 3

The code below uses two variables. What is its running time?

```
1   /* Assume A and B are both arrays.*/
2   for i from 0 to A.length {
3       int j = 0;
4       while (a[i] != b[j]) {
5           print a[i]
6           j = j + 1
7       }
8   }
```

This is said to be O(a*b), where a is the length of A and b is the length of B. Although the inner while loop may sometimes terminate early (having found a[i]), the expected case is that it will iterate through roughly all of B.

CrackingThePMInterview.com   **295**

Example 4

Here is a more challenging example.

```
1   int i = N;
2   while i >= 1 {
3      print i
4      i = i / 2
5   }
```

We need to think about what this for loop will do. This for loop will do something (print a value) and then continuously divide by 2 until it gets below 1.

How many times can we divide N by 2 until we get below 1 and the while loop terminates? If we approached this in reverse, we could say: how many times can we multiply 1 by 2 until we get to N? This would be the value x, where $2^x = n$. This for loop, therefore, iterates x times.

Now we just need to solve for x:

```
2ˣ = n
log(2ˣ) = log(n)
x log(2) = log(n)
x = log(n) / log(2)
```

So, this code operates in $O(\log(n))$ time.

This is a good thing to remember: if something continuously divides in half, it is $O(\log(N))$ time.

Recursion

If a function can call other functions, then it can call itself. This is recursion.

Recursion can be a useful strategy to solve a large number of problems. It works well when the solution to a problem can be defined in terms of the solutions to subproblems.

For example, consider the factorial problem. What is n! (n factorial)? n! is n * (n-1) * (n-2) * ... * 1. We could also say that n! is n * (n-1)!.

This leads to an extremely short bit of code to compute n!.

```
1   int factorial(int n) {
2      if (n == 0 or n == 1) { /* base case */
3         return 1;
4      } else {
5         return n * factorial(n-1);
6      }
7   }
```

The base case (or terminating condition) is extremely important. Without it, the function would run forever.

Here's another example of a recursive function. This computes the nth fibonacci number. As you may recall, the nth fibonacci number, f(n), is f(n-1) + f(n-2).

```
1   int fibonacci(int n) {
2       if (n == 0) {
3           return 0;
4       } else if (n == 1) {
5           return 1;
6       } else {
7           return fibonacci(n-1) + fibonacci(n-2);
8       }
9   }
```

This is a natural function to implement recursively, as the nth fibonacci numbers are defined by their smaller problems.

Memory Usage

Any problem that can be solved recursively can also be solved iteratively (non-recursively), although sometimes doing so is much more complicated. However, recursion comes with a drawback, which is memory usage.

Recall this example:

```
1   int factorial(int n) {
2       if (n == 1) { /* base case */
3           return 1;
4       } else {
5           return n * factorial(n-1);
6       }
7   }
```

This takes $O(N)$ time, and will on any solution. What is its memory usage? Its memory usage will be $O(N)$ too (assuming no fancy optimizations by the compiler).

The method factorial(n) calls factorial(n-1), which calls factorial(n-2), and so on. Note that factorial(n) does not complete until factorial(n-1) completes, which in turn doesn't complete until factorial(n-2).

Therefore, at one point in time, we have n functions in operation at once, on the "call stack."

```
    factorial(0)
    factorial(1)
```

```
...
factorial(n-1)
factorial(n)
```

Each one of those takes up some memory. Therefore, at one point in time, n chunks of memory are being used. This means that this program, when implemented recursively, is $O(N)$ time and $O(N)$ memory.

This is the drawback of recursion: the recursive calls take up memory.

How You Are Evaluated

Some coding and algorithm problems can be quite straightforward, but a good number are actually very challenging. They're challenging for good reason: easy problems wouldn't help interviewers distinguish great candidates from just so-so candidates.

What many candidates don't realize though is that you aren't expected to solve a tough problem immediately. It's great if you can, but it's often not realistic. A tough problem can often take the entire interview to solve, even with some guidance from the interviewer.

Evaluation of your performance is not about whether you got the problem "right" or not. This has little meaning. Rather, the evaluation is a lot more qualitative and, frankly, subjective:

- How willing were you to solve the problem? If you get scared and give up, that's a red flag. Interviewers want candidates who are excited about solving hard problems. They tend to make good employees.

- How quickly did you solve it?

- How optimal was your algorithm?

- How did you approach the problem?

- How much help did you need?

- How clean was your code?

- How was your communication in discussing the problem? How did you react to feedback and guidance from the interviewer?

None of these aspects are yes/no decisions.

An interviewer is not given metrics to decide how quick is "quick" on a problem, or how many bugs constitutes "buggy." How then does she determine how you did?

She determines performance by comparing you—indirectly—to other candidates. That is, the first time she asks the question she doesn't have a good feel for whether X minutes is fast or slow. As she asks more and more people the same question, it starts becoming clearer. When other people typically take 20 - 30 minutes to solve a question and you get it in 10 minutes, she knows that you solved it quickly.

Because the evaluation is relative, it's also extremely difficult to judge by yourself how you did in an interview. You may feel you struggled, but you don't know how much other candidates struggled.

How To Approach

Coding and algorithm questions are designed to test your problem-solving skills. Therefore, you want to show the interviewer how you're approaching the problem.

The following approach works well:

1. **Clarify the Problem:** Make sure you understand what the problem is asking. Ask questions to verify any assumptions. For example, if it's a binary tree, is it a binary search tree? Is it balanced? You might even want to repeat the problem back in your own words.

2. **Go to the Whiteboard:** When you hear a problem, go to the whiteboard and create an example for this problem. Your example should be general enough to help you solve the problem and should avoid special cases.

3. **Talk Out Loud:** Talk out loud and brainstorm solutions with your interviewer. If you can think of a brute-force solution, but you don't think it's good enough, go ahead and explain that solution to your interviewer. At least it will give you a jumping-off point for solving the problem.

4. **Think Critically:** Once you've come up with an algorithm, think through whether it really works. What is its big O time? Can you do better? Does it actually solve the problem? Are there any cases where it will fail?

5. **Code, Slowly and Methodically:** Once you and your interviewer are comfortable with the code, go ahead and start coding on the whiteboard. If it helps, you can write out some pseudocode first. Make sure when you're coding that you really, truly understand what you're doing. If you get confused while you're coding, take a step back and think through your algorithm. Rushing will not help you do better.

6. **Test and Fix:** Just because there's no computer doesn't mean you don't test. You must test your code. In this case, you'll walk through your code with the edge cases and normal cases. When you find bugs—and you will (bug-free code is unusual in an interview)—think through what caused the bug, and then carefully fix it.

Observe that the coding part is Step 5, not Step 1. Do not just get up to the whiteboard and start coding once you hear a problem. Take your time to brainstorm a solution with your interviewer. The algorithm part often takes up most of the interview, and sometimes even the entire interview.

Remember: if you're struggling to solve a problem, that's normal. These problems are designed to be difficult.

Developing an Algorithm

As we've said, many problems are difficult. How then do you come up with an algorithm? Here are a few strategies that work well:

- **Use an Example:** When you get a problem, don't just sit there in your chair and try to solve it in your head. Get up, go to the whiteboard, and sketch out an example. For example, if you wanted to figure out how to merge two sorted arrays, draw an example of two specific arrays, like {1, 5, 8, 9} and {3, 5, 7, 10, 12}, and walk through how you would merge these. Use specific values (not just variable names like a1 and a2) and avoid special cases (such as the arrays having no elements in common).

- **Optimize the Brute Force:** There is no shame at all in starting with a brute force solution or a naive solution. It gives you a good starting point from which you can optimize. If the brute force is slow, think about *why* it's slow. Which steps of the algorithm are the biggest time hogs? Focus on optimizing those.

- **Solve for Base Case:** Sometimes, it's easy to solve a problem for small cases. Try solving it for 0, then 1, then 2, and so on. Can you see a pattern? Or, can you build the answers for these larger cases using the answers for a prior one? For example, if you're trying to compute all subsets of {a, b, c}, you might be able to use subsets of {a, b} to do it.

- **Think about Similar Problems:** If the problem sounds similar to others you've heard before, see if a similar approach will work for this problem. The more practice you do on solving technical questions, the easier it will be to come up with an algorithm.

- **Simplify and Tweak:** Interview problems come with certain constraints, in the size of the input, the type of input, the ranges, or another factor. Try tweaking or simplifying the problem in some way to see if you can solve this alternate version.

- **Record Your Insights:** Some problems have key "insights" that you might discover during your problem solving. For example, maybe you're looking for a value in a tree and you realize that it has to be on the right side of a left subtree; remember that insight.

Whatever you do, approach the problem energetically. Don't be discouraged when you struggle, and don't give up. Interviewers want to see that you'll push your way through problems.

Additional Questions

16.19 Design an algorithm and write code to find all solutions to the equation $a^3 + b^3 = c^3 + d^3$ where a, b, c, and d are positive integers less than 1000. If you wish, you can print only "interesting" solutions. That is, you can ignore solutions of the form $x^3 + y^3 = x^3 + y^3$ and solutions that are simple permutations of other solutions (swapping left and right hand sides, swapping a and b, swapping c and d). For example, if you were printing all solutions less than 20, you could choose to print only $2^3 + 16^3 = 9^3 + 15^3$ and $1^3 + 12^3 = 9^3 + 10^3$.

pg 326

16.20 Given a string, print all permutations of that string. You can assume the word does not have any duplicate characters.

pg 330

16.21 In a group of people, a person is called a "celebrity" if everyone knows them but they know no one else. You are given a function knows(a, b) which tells you if person a knows person b. Design an algorithm to find the celebrity (if one exists).

For simplicity, you can assume that everyone is given a label from 0 to N-1. You need to implement a function int findCelebrity(int N).

Observe that:

(1) There can only be one celebrity at most (due to the definition of a celebrity).

(2) The knows function is the *only* way to look up who knows who.

pg 332

16.22 You have an NxN matrix of characters and a list of valid words (provided in any format you wish). A word can be formed by starting with any character and then moving up, down, left, or right. Words do not have to be in a straight line (PACKING is a word below). You cannot reuse a letter for the same word, so GOING (in the grid below) would not be a word since it reuses the G. Design an algorithm and write code to print all valid words.

```
L I G O
E P N I
N A C K
S M A R
```

pg 334

16.23 Given an array of integers (with both positive and negative values), find the contiguous sequence with the largest sum. Return just the sum.

Example:

```
Input:  2, -8, 3, -2, 4, -10
Output: 5 (i.e., {3, -2, 4})
```

pg 337

Solutions

All solutions will be implemented with Java. If you don't know Java, that's okay. We'll try to keep our code free of complex Java syntax so that you can focus on the main algorithm.

16.1   *Given a sorted array of positive integers with an empty spot (zero) at the end, insert an element in sorted order.*

pg 282

We can imagine that our array looks something like this (with a blank spot at the end):

    1 4 7 8 9 _

If we need to insert an element like 6, we can't just insert it at the end. We are supposed to put it in order.

    1 4 6 7 8 9

This requires "shifting" all elements down to make space for 6 and then inserting it.

There are two ways of approaching this problem.

Approach 1: Shift From Back, Then Insert

The first approach is to shift all the elements over and then insert the value x. We have to be careful though to not overwrite values as we're inserting.

Instead of shifting from the front, we can shift from the back moving forwards.

    1 4 7 8 9 _

We would first copy 9 into the empty spot. Then 8 into where 9 was. Then 7 into where 8 was, and so on. When we find the appropriate spot for x, we stop and insert x.

```
1   boolean insert(int[] array, int x) {
2     /* Make sure input is valid.*/
3     if (array[array.length - 1] != 0 || x <= 0) {
4       return false;
5     }
6
7     /* Start from last non-blank element, moving left and copying
8      * elements one by one. Stop when we've found the right spot
9      * for x or when we've hit the beginning of the array.*/
10    int index = array.length - 2; // start from 2nd to last
11    while (index >= 0 && array[index] > x) {
```

```
12      array[index + 1] = array[index]; // shift over by one
13      index = index - 1; // move to next element
14    }
15
16    /* Insert element wherever the above loop stopped.*/
17    array[index + 1] = x;
18
19    return true;
20  }
```

We return true if we could insert the element or false if there was an error.

Approach 2: Swap Elements Moving Forward

Alternatively, we could iterate forwards through the array. For the initial elements in the array (the ones that are less than x), we don't do anything. Those won't be moved.

However, when we find where x should be inserted, we swap x and the current element in the array. The value of x will now equal the old element in the array.

When we get to the next element, we want to swap x for that value. We continue doing this for each element in the array until we get to the end.

```
insert 6 into 2, 3, 7, 8, 9, _
set x = 6
start i at A[0]
move i to A[1]
move i to A[2]
swap A[2] and x.
    A = {2, 3, 6, 8, 9, _}
    x = 7
swap A[3] and x.
    A = {2, 3, 6, 7, 9, _}
    x = 8
swap A[4] and x.
    A = {2, 3, 6, 7, 8, _}
    x = 9
swap A[5] and x.
    A = {2, 3, 6, 7, 8, 9}
    x = _
```

The following code implements this algorithm.

```
1  boolean insert(int[] array, int x) {
2    /* Make sure input is valid.*/
3    if (array[array.length - 1] != 0 || x <= 0) {
4      return false;
5    }
6
7    for (int i = 0; i < array.length; i++) {
```

```
8      if (x < array[i] || array[i] == 0) {
9          /* Swap x and array[i].*/
10         int temp = array[i];
11         array[i] = x;
12         x = temp;
13     }
14  }
15
16    return true;
17 }
```

Note that once the if statement on line 8 becomes true, it will always be true.

Both algorithms will take O(N) time.

16.2   *Reverse the order of elements in an array (without creating a new array).*

pg 282

At first glance, we might want to just create a second array, iterate over the elements in order, and insert them in reverse order into the new array. Unfortunately, the question says to not create a second array.

Let's look at any example.

```
Original:   0, 1, 2, 3, 4, 5, 6
Reversed:   6, 5, 4, 3, 2, 1, 0
```

You might notice that by reversing the array, we're putting the 0 where the 6 is and the 6 where the 0 is. Likewise, the 5 and the 1 are put in each other's places. That is, we're swapping values!

Rather than create a second array, we can iterate through the array, swapping the left values with the corresponding values on the right. We only need to iterate through the left half of the array, since the right half of the array will have been taken care of already.

```
1   void reverse(int[] array) {
2       int midpoint = array.length / 2;
3       for (int i = 0; i < midpoint; i++) {
4          /* Get corresponding index on right side.*/
5          int otherside = array.length - 1 - i;
6
7          /* Swap left and right values.*/
8          int temp = array[otherside];
9          array[otherside] = array[i];
10         array[i] = temp;
11     }
12 }
```

Be very careful with the arithmetic on lines 2 and 5. Those are the sorts of things you should double and triple check in an interview.

Both algorithms will take O(N) time.

16.3   *Given two lists (A and B) of unique strings, write a program to determine if A is a subset of B. That is, check if all the elements from A are contained in B.*

pg 283

We're told that the two lists contain unique strings, so we only need to check if all the elements in one list are contained in the other.

Approach 1: Brute Force

We can approach this by "brute force." For each element in A, check if it is in B.

As soon as we find an element a in A which is not in B, we can return `false` because we know A is not a subset. If we reach the end of A and haven't returned yet, then we know we were able to find every element. We return `true`.

```
1   boolean isSubsetBruteForce(String[] bigger, String[] smaller) {
2      for (String s : smaller) {
3         boolean found = false;
4         for (String b : bigger) {
5            if (s.equals(b)) { // found element
6               found = true;
7               break;
8            }
9         }
10        if (!found) { // s wasn't found -> not subset
11           return false;
12        }
13     }
14     return true; // all elements found
15  }
```

This algorithm takes O(a*b) time, where a is the length of A and b is the length of B.

Approach 2: Hashtable

The reason that the earlier approach is so slow is that we have to search through B for each element. Wouldn't it be nice if we could just look up if an element is in B?

We can! This is what a hashtable allows us to do. We can build a hashtable of all the elements in B. Then, when we want to look up if an element is in B, we just use that hashtable.

```
1  boolean isSubset(String[] bigger, String[] smaller) {
2     Hashtable<String, Boolean> hash =
3        new Hashtable<String, Boolean>();
4
5     /* Record all the elements in the bigger list.*/
6     for (String b : bigger) {
7        hash.put(b, true);
8     }
9
10    /* Check if the bigger hashtable contains all the strings.*/
11    for (String s : smaller) {
12       if (!hash.containsKey(s) || hash.get(s) != true) {
13          return false;
14       }
15    }
16    return true;
17 }
```

This algorithm takes O(a+b) time, where a is the length of A and b is the length of B. It takes O(b) additional memory to hold the hashtable.

16.4 *You are given a two-dimensional array of sales data where the first column is a product ID and the second column is the quantity. Write a function to take this list of data and return a new two-dimensional array with the total sales for each product ID.*

Example:

Input:

211,4
262,3
211,5
216,6

Output:

211,9
262,3
216,6

pg 283

The output for this method needs to be a list of product IDs and their total counts. We can do this in a straightforward manner by using a hashtable.

We iterate through the list of (productID, quantity) pairs. For each value, we increment its entry in the hashtable or insert it if it's not already in there. Finally, we convert the hashtable back into an array.

```
1  int[][] totalSales(int[][] data) {
2     Hashtable<Integer, Integer> hash =
```

```
3          new Hashtable<Integer, Integer>();
4
5      /* Compute total sales of each product.*/
6      for (int i = 0; i < data.length; i++) {
7          int productId = data[i][0];
8          int quantity = data[i][1];
9          if (hash.containsKey(productId)) {
10             quantity = quantity + hash.get(productId);
11         }
12         hash.put(productId, quantity);
13     }
14
15     /* Convert hashtable back to array.*/
16     int[][] totals = new int[hash.keySet().size()][2];
17     int index = 0;
18     for (int key : hash.keySet()) {
19         totals[index][0] = key;
20         totals[index][1] = hash.get(key);
21         index = index + 1;
22     }
23     return totals;
24 }
```

If you don't know the specific commands for things like keySet and contain-sKey, don't worry. Your interviewer shouldn't care about things like this. The important thing is that you know how to translate an approach into something that resembles workable code.

This algorithm takes O(N) time, where N is the number of lines in the input.

16.5 *Insert an element into a binary search tree (in order). You may assume that the binary search tree contains integers.*

pg 284

This is a straightforward question that follows from the definition of a binary search tree.

In a binary search tree, lesser values are put on the left of a node and greater values are put on the right.

The easiest way to implement this is recursively. Start with the root and compare the value you want to insert, x. If x is less than the root, then call insert on the root.left. When x is greater than the root, then call insert on the right side. Repeat this until you don't have a left or right child. Insert x there.

```
1   boolean insert(TreeNode root, int data) {
2       if (root == null) return false; // failure
3
```

```
4      if (data <= root.data) {
5          if (root.left == null) { // found spot
6              root.left = new TreeNode(data); // insert
7          } else {
8              return insert(root.left, data); // recurse
9          }
10     } else {
11         if (root.right == null) { // found spot
12             root.right = new TreeNode(data); // insert;
13         } else {
14             return insert(root.right, data); // recurse
15         }
16     }
17     return true; // success
18  }
```

The time to insert a node will depend on the height of the tree. If the tree is relatively balanced, it should have height O(log N) where N is the number of nodes in the tree. However, if the tree is very imbalanced (for example, basically a straight line down of nodes all on one side), the height could be as much as N.

16.6 *Given a binary search tree which contains integers as values, calculate the sum of all the numbers.*

pg 284

If we approach problems from the right perspective, some problems are surprisingly simple. In this case, the "right" perspective means recursively.

Suppose we want to compute the sum of the nodes in a tree like this:

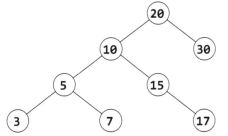

We could traverse the tree, collapse it into an array, and then compute the sum of those values. That's a lot more complicated than is necessary.

The simpler way is to think about the problem in terms of its subproblems. The sum of the entire tree is going to be the sum of the left subtree + sum of the right subtree + sum of the root.

```
sum(tree_at_20) =
        sum(tree_at_10)
```

```
    + sum(tree_at_30)
    + value_at_node_20
```

Getting the sum at node 10 can then be defined in terms of its subproblems.

```
sum(tree_at_10) =
     sum(tree_at_5)
   + sum(tree_at_15)
   + value_at_node_10
```

We can almost directly translate this into code.

```
1   int sum(TreeNode root) {
2       if (root == null) {
3           return 0;
4       }
5       return root.data + sum(root.left) + sum(root.right);
6   }
```

If we hit the end of a path (a null node), we return 0. If you prefer, we could instead do this:

```
1   int sum(TreeNode root) {
2       if (root == null) {
3           return 0;
4       }
5
6       int total = root.data;
7       if (root.left != null) {
8           total += sum(root.left);
9       }
10      if (root.right != null) {
11          total += sum(root.right);
12      }
13      return total;
14  }
```

Regardless of which code you use, the runtime will be O(N), where N is the number of nodes in the tree.

One way to see the runtime is to realize that sum will be called exactly once for each node in the tree. If there are O(N) calls to sum, then the runtime should be O(N).

16.7   *Insert a node into a sorted linked list (in order). (Don't forget about what happens when the new element is at the start or end!)*

To insert a number in order into a linked list, we first need to find the right place to insert the node. Then, we need to actually insert it.

The tricky bit is figuring out how to handle inserting a node into the front of the linked list.

Imagine we call an `insertInOrder` method that looks like this, and it (for this particular case) needs to insert n into the front of the linked list:

```
void insertInOrder(LinkedListNode nd, int value)
```

Just inserting node n and having n.next point to nd is not enough. Whoever is using the linked list doesn't know that the *real* head of the linked list has been updated from nd to n. They only have a reference to nd.

Therefore, in an insert method, you need to return the new head of the linked list. Most of the time, the head will be the same as it was before you called insert. Sometimes it will change though, and you need to notify the caller of this.

```
1   LinkedListNode insert(LinkedListNode head, int data) {
2       /* Create new node.*/
3       LinkedListNode node = new LinkedListNode(data);
4
5       /* If being insert into front of linked list, then insert
6        * node and return new head.*/
7       if (head == null || data < head.data) {
8           node.next = head;
9           return node;
10      }
11
12      /* Find the right spot (in order) to insert the node by
13       * traversing through the list.*/
14      LinkedListNode current = head;
15      while (current.next != null && data > current.next.data) {
16          current = current.next;
17      }
18
19      /* Insert node.*/
20      node.next = current.next;
21      current.next = node;
22
23      /* Return old head. It hasn't moved.*/
24      return head;
25  }
```

This algorithm takes O(N) time, where N is the number of nodes.

16.8   *"Sort" a linked list that contains just 0s and 1s. That is, modify the list such that all 0s come before all 1s.*

pg 285

There are many ways of doing this problem.

Approach 1: Build Two Linked Lists

One of the simplest ways is to build a "zeros list" and a "ones list" and then join them at the end.

```
1   LinkedListNode sort(LinkedListNode head) {
2       LinkedListNode zeroHead = null;
3       LinkedListNode zeroTail = null;
4       LinkedListNode oneHead = null;
5       LinkedListNode oneTail = null;
6
7       LinkedListNode n = head;
8       while (n != null) {
9           LinkedListNode next = n.next;
10          n.next = null;
11          if (n.data == 0) {
12              /* Add to end of zeros list.*/
13              if (zeroHead == null) {
14                  zeroHead = n;
15              } else {
16                  zeroTail.next = n;
17              }
18              zeroTail = n;
19          } else {
20              /* Add to end of ones list.*/
21              if (oneHead == null) {
22                  oneHead = n;
23              } else {
24                  oneTail.next = n;
25              }
26              oneTail = n;
27          }
28          n = next;
29      }
30
31      /* Join lists and return.*/
32      if (zeroTail == null) {
33          oneTail.next = null;
34          return oneTail;
35      } else {
36          zeroTail.next = oneHead;
37          return zeroHead;
38      }
39  }
```

Observe that we need to return the new head of the linked list, as it might have changed.

Approach 2: Grow Left and Right

The prior approach gets lengthy because we need to keep track of the front and back of two different lists, which requires constantly updating four different variables.

Four variables aren't actually necessary though. All the problem has asked us to do is to put all the 0s before the 1s. We don't have to keep the nodes in the same order that they were in originally.

Therefore, we can instead just keep track of the head and tail of the (new) linked list. When we get a new 0, we insert it at the very front. When we get a new 1, we insert it at the very end. This will keep all 0s before every 1.

```
1   LinkedListNode sort(LinkedListNode n) {
2       LinkedListNode head = n;
3       LinkedListNode tail = n;
4       n = n.next; // start with second element
5
6       while (n != null) {
7           LinkedListNode next = n.next;
8           if (n.data == 0) { // 0 -> add to front
9               n.next = head;
10              head = n;
11          } else { // 1 -> add to tail
12              tail.next = n;
13              tail = n;
14          }
15          n = next;
16      }
17      tail.next = null; // ensure tail doesn't point anywhere
18
19      return head;
20  }
```

We again need to return the new head since it might have changed.

Approach 3: Count the Zeros

We're not actually required to use the same actual objects that we were given. If we moved values, instead of nodes, that would fit the problem requirements.

Therefore, we can just iterate through the linked list once, counting the number of 0s. Then, we iterate through it again, setting the first k values to 0 and the rest to 1.

```
1   void sort(LinkedListNode head) {
2       int k = 0;
3
```

```
4      /* Count the number of 0s.*/
5      LinkedListNode n = head;
6      while (n != null) {
7        if (n.data == 0) {
8          k = k + 1;
9        }
10       n = n.next;
11     }
12
13     /* Set first k values to 0.*/
14     n = head;
15     while (n != null) {
16       if (k > 0) {
17         n.data = 0;
18         k = k - 1;
19       } else {
20         n.data = 1;
21       }
22       n = n.next;
23     }
24 }
```

In this approach, we're moving values, not nodes. The actual reference to the head won't change, so we don't need to return anything.

Approach 4: Swap the Values

Since we only need to move the values, we can also just iterate through the linked list, swapping the 0s and 1s as we find them.

This approach works by two pointers, p and q. The p pointer looks for 1s and the q pointer looks for 0s. When they find their values, they swap.

1. Start p at head.

    ```
    0->0->0->1->1->0->1->0->1->0
    p
    ```

2. Move p to first 1.

    ```
    0->0->0->1->1->0->1->0->1->0
             p
    ```

3. Start q at p.next.

    ```
    0->0->0->1->1->0->1->0->1->0
             p  q
    ```

4. Move q to next 0.

    ```
    0->0->0->1->1->0->1->0->1->0
             p     q
    ```

5. Swap values at p and q.

```
0->0->0->0->1->1->1->0->1->0
         p    q
```

6. Repeat at step 4:

```
0->0->0->0->1->1->1->0->1->0
              p  q          // move p to next 1
0->0->0->0->1->1->1->0->1->0
              p       q     // move q to next 0
0->0->0->0->0->1->1->1->1->0
              p       q     // swap
0->0->0->0->0->1->1->1->1->0
                 p    q     // move p to next 1
0->0->0->0->0->1->1->1->1->0
                 p       q  // move q to next 0
0->0->0->0->0->0->1->1->1->1
                 p       q  // swap
```

In other words, p is always pointing to the first 1 and q is always pointing to the first out of place 0 (which is the first 0 after p). Whenever q finds a 0, we know the 0 is out of place. We swap its value with p and move p to the next node.

This approach might be the least intuitive for some people, but—with the use of a helper function—it leads to fairly short code.

```
1   void sort(LinkedListNode head) {
2      LinkedListNode p = nextNodeWithVal(head, 1);    // find first
1
3      LinkedListNode q = nextNodeWithVal(p.next, 0); // find next 0
4
5      while (p != null && q != null) {
6         q.data = 1; // swap
7         p.data = 0;
8         p = nextNodeWithVal(p, 1); // find next 1
9         q = nextNodeWithVal(q, 0); // find next 0
10     }
11  }
12
13  LinkedListNode nextNodeWithVal(LinkedListNode n, int data) {
14     while (n != null && n.data != data) {
15        n = n.next;
16     }
17     return n;
18  }
```

These are just four ways of solving the problem, but there are many other ways too. These four approaches are all O(N).

16.9   *Write a function which takes a stack as input and returns a new stack which has the elements reversed.*

pg 285

The most straightforward way to do this is to just create a new stack and pop the elements from the first stack onto the second. This will put the top element from the original stack on the bottom of the new stack.

```
1   Stack<Integer> reverse(Stack<Integer> stack) {
2       Stack<Integer> reversed = new Stack<Integer>();
3       while (!stack.isEmpty()) {
4           reversed.push(stack.pop());
5       }
6       return reversed;
7   }
```

The only problem with this is that our original stack gets completely emptied in the process. If this is a problem (ask your interviewer!), then you can use an additional stack to hold all the popped values.

We push the popped values onto both the temp stack and the reversed stack. (These stacks will have the same elements in the same—reversed—order.) Once we're done popping the elements from stack, we push them back from temp onto stack.

```
1   Stack<Integer> reverseWithoutDestroying(Stack<Integer> stack) {
2       Stack<Integer> reversed = new Stack<Integer>();
3       Stack<Integer> temp = new Stack<Integer>();
4       while (!stack.isEmpty()) {
5           int x = stack.pop();
6           reversed.push(x);
7           temp.push(x);
8       }
9
10      while (!temp.isEmpty()) {
11          stack.push(temp.pop());
12      }
13
14      return reversed;
15  }
```

Both approaches will have O(N) runtime. The second one will go through two passes instead of one, but constants don't affect the big O time. This may seem surprising to some people, but remember: big O is not an expression of how many seconds something actually takes. It expresses how the time scales (in this case, linear) as the size of the input gets longer and longer.

16.10 *Write a function which removes all the even numbers from a stack. You should return the original stack, not a new one.*

pg 285

For this problem, we can rely on the same instinct as the second approach from the prior problem: reversing something twice puts the elements back in their original order.

We can just pop the stack, element by element. If the element is odd (that is, not even), push it onto a new, temporary stack. Then, once we're all done, push them back onto their original stack.

```
1   void removeEvens(Stack<Integer> stack) {
2       Stack<Integer> temp = new Stack<Integer>();
3       while (!stack.isEmpty()) {
4           int x = stack.pop();
5           /* Push evens onto new stack.*/
6           if (x % 2 != 0) {
7               temp.push(x);
8           }
9       }
10
11      /* Return odds to the stack.*/
12      while (!temp.isEmpty()) {
13          stack.push(temp.pop());
14      }
15  }
```

This algorithm will take O(N) time. Observe that, since you have to go through every element, you can't solve the problem any faster than this.

16.11 *Write a function to check if two queues are identical (same values in the same order). It's okay to modify/destroy the two queues.*

pg 286

We are allowed to modify the two queues, which should give us a clue that we need to do just that.

We can repeatedly remove the front of each linked list and compare the values. If the values are not equal, then we immediately return false.

What happens when one list is emptied? That depends. If both lists are empty, then we know the linked lists are identical (nothing has failed yet). However, if only one list is empty and the other is not, then we know the lists were of different sizes. After all, we're removing the elements in the same order.

```
1   boolean isEqual(Queue<Integer> one, Queue<Integer> two) {
2       /* Remove elements one by one and check that they're equal.*/
```

```
3      while (!one.isEmpty() && !two.isEmpty()) {
4          int oneHead = one.remove();
5          int twoHead = two.remove();
6          if (oneHead != twoHead) {
7              return false;
8          }
9      }
10
11     /* We exited because one list was emptied. If the other list
12      * has elements left in it, then they must have been of
13      * different sizes. */
14     if (!one.isEmpty() || !two.isEmpty()) {
15         return false;
16     }
17
18     return true;
19 }
```

This algorithm takes O(N) time, where N is the length of the smaller list. Why the smaller? Because we exit as soon as *either* list is empty. That will happen to the smaller list first. It doesn't matter how big the bigger list is; it won't affect the runtime.

16.12 *Write a function to remove the 13th element from a queue (but keep all the other elements in place and in the same order).*

pg 286

The approach to this problem depends on what you assume the queue data structure supports.

If we get access to a Node class, then this is fairly easy. We just iterate through the nodes and delete it when we get to the 13th.

However, if it's a true Queue class, we don't necessarily have access to the nodes like this. We may only have an add (to the back of the list) and remove (from the front of the list) method.

We could create a second list object, but this isn't actually necessary. Observe that if we continuously remove elements from the front and add them to the back, we'll wind up with the exact same list.

To remove the 13th element, we can therefore just remove each element and re-add it—skipping the 13th element.

We've implemented this code using a variable k, rather than hard coding the number 13. This is generally a good coding practice.

```
1   boolean remove(Queue<Integer> queue, int k) {
2     if (k < 0 || k >= queue.size()) {
3       return false;
4     }
5     int size = queue.size();
6     for (int i = 0; i < size; i++) {
7       int head = queue.remove();
8       if (i != k) { /* everything but the kth element */
9         /* remove from front and add to back */
10        queue.add(head);
11      }
12    }
13    return true;
14  }
```

This algorithm takes O(N) time, where N is the number of nodes.

16.13 *Given two sorted arrays, write a function to merge them in sorted order into a new array.*

pg 287

The most efficient way to tackle this is to use the fact that the arrays are sorted. We can merge them by taking successive elements repeatedly until we reach the end of both arrays. We maintain pointers to where we are in each array so that we can just easily move onto the next array.

Let's take an example of two arrays.

```
    A: 1 5 8 9     B: 2 4 9 10 12
```

We'll start off with the p and q pointers at the beginning of the two arrays:

```
    A: 1 5 8 9     B: 2 4 9 10 12
       p              q
```

A[p] is smaller than A[q], so we put A[p] into our result array. We then move p to the next value.

```
    A: 1 5 8 9     B: 2 4 9 10 12     Result: 1
       p              q                       i
```

We compare A[p] and A[q] again, putting the smaller element into the resulting array. We also need to keep track of where we are in the result array. We repeat this process until we are done with both arrays.

```
    A: 1 5 8 9     B: 2 4 9 10 12     Result: 1 2
         p            q                         i
    A: 1 5 8 9     B: 2 4 9 10 12     Result: 1 2 4
         p              q                         i
    A: 1 5 8 9     B: 2 4 9 10 12     Result: 1 2 4 5
           p            q                           i
```

```
A: 1 5 8 9        B: 2 4 9 10 12       Result: 1 2 4 5 8
         p                 q                              i
A: 1 5 8 9        B: 2 4 9 10 12       Result: 1 2 4 5 8 9
                           q                                i
A: 1 5 8 9        B: 2 4 9 10 12       Result: 1 2 4 5 8 9 9
                             q                                  i
A: 1 5 8 9        B: 2 4 9 10 12       Result: 1 2 4 5 8 9 9 10
                               q                                  i
A: 1 5 8 9        B: 2 4 9 10 12       Result: 1 2 4 5 8 9 9 10 12
                                                                    i
```

In an interview, it's useful to walk through the example in this detail to reduce the number of mistakes you make.

```
1   int[] mergeIntoNew(int[] A, int[] B) {
2       int p = 0;
3       int q = 0;
4       int index = 0;
5       int[] merged = new int[A.length + B.length];
6
7       while (p < A.length || q < B.length) {
8           if (q >= B.length || A[p] <= B[q]) {
9               merged[index] = A[p];
10              p = p + 1;
11          } else {
12              merged[index] = B[q];
13              q = q + 1;
14          }
15          index = index + 1;
16      }
17
18      return merged;
19  }
```

If you wanted to reduce the number of pointers we have, we can remove the index variable. It will always be equal to p + q.

This code takes O(M+N) time, where M is the length of the first array and N is the length of the second.

16.14 *Implement insertion sort.*

pg 301

Insertion sort operates by iterating through the array, inserting each element in order on the element's left side.

We can most cleanly implement this as two different functions.

The first function performs the overall algorithm: pick up an element, insert it in

---

order, pick up the next one, and so on.

```
1   void insertionSort(int[] array) {
2       /* Pick up elements starting from the left and insert them into
3        * the left in order */
4       for (int i = 1; i < array.length; i++) {
5           insertInOrder(array, i);
6       }
7   }
```

Observe that our for loop starts at 1 instead of 0. This is because the 0th element can never be out of order, by itself. (A single element subarray is always sorted.)

Now we just need to implement a method that will take an element A[k] and insert it in order into the elements to the left of it (provided those are sorted).

To insert A[k] in order, we will need to shift each element over by one, until we find the right spot for the element.

```
1   void insertInOrder(int[] array, int index) {
2       int x = array[index]; // store element into temporary value
3       index = index - 1;
4       while (index >= 0 && array[index] > x) {
5           array[index + 1] = array[index]; // shift over by one
6           index = index - 1;
7       }
8       array[index + 1] = x; // insert element
9   }
```

This algorithm will take $O(N^2)$ time.

16.15 *Implement binary search. That is, given a sorted array of integers and a value, find the location of that value.*

pg 289

Binary search works by repeatedly "halving" the array into subarrays. In the first iteration, we compare the value x to the midpoint and learn whether x will be in the left half or the right half. Then, we repeat this step with this new subarray: is x found on the left half of it (the new subarray) or the right?

We can implement this either recursively or iteratively (non-recursively). We'll start with the recursive solution since it's more intuitive for most people.

```
1   int search(int[] array, int x) {
2       return search(array, x, 0, array.length - 1);
3   }
4
5   int search(int[] array, int x, int left, int right) {
6       if (right < left) { // Not found
```

```
7          return -1;
8      }
9
10     int middle = (right + left) / 2;
11     if (x == array[middle]) {
12         return middle;
13     } else if (x < array[middle]) { // x is on left hand
14         return search(array, x, left, middle - 1);
15     } else { // x is on right hand
16         return search(array, x, middle + 1, right);
17     }
18 }
```

For the iterative solution, we take a very similar approach.

```
1   int search(int[] array, int x) {
2       int left = 0;
3       int right = array.length - 1;
4       while (left <= right) {
5           int middle = (right + left) / 2;
6           if (x == array[middle]) {
7               return middle;
8           }
9           if (x < array[middle]) {
10              right = middle - 1;
11          } else {
12              left = middle + 1;
13          }
14      }
15      return -1;
16 }
```

A good exercise is to think about how different bits of logic from the recursive solution translates to the iterative solution. For example, what happened to the check on line 6 of the recursive solution?

16.16 *You are given an integer array which was sorted, but then rotated. It contains all distinct elements. Find the minimum value. For example, the array might be 6, 8, 9, 11, 15, 20, 3, 4, 5. The minimum value would obviously be 3.*

pg 288

A brute force solution would be to just iterate through the array and look for the minimum value. We can guess that this isn't what the interviewer is looking for though, since it doesn't use the sorting information.

To come up with a more optimal solution, we probably want to use the information we're given—the array is "sorted," but rotated.

Since the array is somewhat sorted, let's think about applying some of the

---

concepts from binary search. Binary search works by looking at the midpoint repeatedly.

In this problem, what does the midpoint tell us? In and of itself, the midpoint being 15 doesn't tell us anything. However, if we know that the left side is 6 and the right side is 5, we can conclude something. Since left > right, we know that the array is out of order. But since left < middle, we know the left is in order but the right is not.

6, _, _, _, 15, _, _, _, 5

From examining the above array, we can determine that the inflection point (which is the minimum element) is on the right half. Our problem is now divided in half.

To find the minimum element, we now just recurse.

20, _, _, 5
20, 3
3

We can implement this recursively. We stop when we find that the left side is less than the right side. This indicates that this portion of the array is in order, and therefore that the left is the smallest element.

```
1   int findMin(int[] array, int left, int right) {
2     /* Items are in order. Therefore, left must be minimum.*/
3     if (array[left] <= array[right]) {
4       return left;
5     }
6
7     /* Find half with minimum element.*/
8     int middle = (right + left) / 2;
9     if (array[left] > array[middle]) {
10      return findMin(array, left, middle);
11    } else { // middle element > right element
12      return findMin(array, middle + 1, right);
13    }
14  }
```

Alternatively, we can implement this algorithm iteratively with a while loop.

```
1   int findMin(int[] array) {
2     int left = 0;
3     int right = array.length - 1;
4     while (array[left] > array[right]) {
5       int middle = (left + right) / 2;
6       if (array[left] > array[middle]) {
7         right = middle;
8       } else {
9         left = middle + 1;
10      }
```

```
11   }
12   return left;
13 }
```

Be very careful in problems like this with your termination and recursion conditions. Think about things like why you make `left = middle + 1` (why the +1?) but `right = middle`. Those are easy places to make mistakes.

16.17 *Using depth-first search, check if a tree contains a value.*

Depth-first search works by checking if a value v is equal to the current node's value. If it is not, then you search each child of the node, one by one.

The difference between depth-first search (DFS) and breadth-first search (BFS) is that in DFS, the entire subtree of a node's child is searched before you move on to any of the node's other children. That is, all of `node.child[0].children` will be searched before you even look at `node.child[1]`.

We can implement this recursively.

```
1   boolean depthFirstSearch(TreeNode node, int x) {
2     if (node == null) {
3       return false;
4     } else if (node.data == x) {
5       return true;
6     } else {
7       return depthFirstSearch(node.left, x) ||
8              depthFirstSearch(node.right, x);
9     }
10 }
```

Because this is a tree, we don't need to be concerned about infinite loops. That is, we don't need to be concerned about traversing through node's children, and node's "grandchildren," and accidentally winding up back at node—to be forever stuck in an infinite loop (yikes!). Trees specifically forbid cycles like this.

If this weren't the case—if we were in a graph instead of a tree—we would have to use an `isVisited` flag to indicate that we've already visited this node.

16.18 *Write the pseudocode for breadth-first search on a binary tree. Try to be as detailed as possible.*

pg 289

To perform breadth-first search, we want to search a node level by level. That is, we want to search each of node's children before we search any of *their* children.

Although breadth-first search is conceptually straightforward (just search a

---

node's children, level by level), implementing it can be a little less intuitive. The main trick to remember is that we need to use a queue.

A queue, as you might recall, is a data structure that allows us to add items on one side and remove them from the other side. It is a "first in, first out" (FIFO) data structure. This enables us to basically flag nodes "as to be processed later."

In BFS, we "visit" a node by comparing the value we're searching for (x) to the current value. If it matches, we're done and can immediately return true. Else, then we add node's children to the end of the queue. We then move on, pulling a node from the other side and searching it.

```
1   boolean searchBFS(TreeNode root, int x) {
2       Queue<TreeNode> queue = new LinkedList<TreeNode>();
3       return searchBFS(root, x, queue);
4   }
5
6   boolean searchBFS(TreeNode root, int x, Queue<TreeNode> queue) {
7       queue.add(root);
8       while (!queue.isEmpty()) {
9           TreeNode node = queue.remove();
10          if (node.data == x) {
11              return true;
12          }
13          if (node.left != null) {
14              queue.add(node.left);
15          }
16          if (node.right != null) {
17              queue.add(node.right);
18          }
19      }
20      return false;
21  }
```

Because this is a tree, we do not need to worry about winding up in a cycle. However, if this were not the case, we would need to use an isVisited flag to ensure we don't revisit the same node.

Breadth-first search takes O(N) time, where N is the number of nodes in the graph (or tree). This is because we might potentially need to search all of the nodes.

16.19 *Design an algorithm and write code to find all solutions to the equation a^3 + $b^3 = c^3 + d^3$ where a, b, c, and d are positive integers less than 1000. If you wish, you can print only "interesting" solutions. That is, you can ignore solutions of the form $x^3 + y^3 = x^3 + y^3$ and solutions that are simple permutations of other solutions (swapping left and right hand sides, swapping a and b, swapping c and d). For example, if you were printing all solutions less than 20, you could choose to print only $2^3 + 16^3 = 9^3 + 15^3$ and $1^3 + 12^3 = 9^3 + 10^3$.*

pg 301

We can start off with a naive solution. We just iterate through all possible values for a, b, c, and d. When they are equal, we can print this set.

```
1   void cubes(int max) {
2       for (int a = 0; a < max; a++) {
3           int acubed = a * a * a;
4           for (int b = 0; b < max; b++) {
5               int bcubed = b * b * b;
6               for (int c = 0; c < max; c++) {
7                   int ccubed = c * c * c;
8                   for (int d = 0; d < max; d++) {
9                       int dcubed = d * d * d;
10                      if (acubed + bcubed == ccubed + dcubed) {
11                          String solution = a + "," + b + "," + c + "," + d;
12                          System.out.println(solution);
13                      }
14                  }
15              }
16          }
17      }
18  }
```

This is a good start. Now, how can we make it faster?

We can get some minor wins by "short circuiting"—i.e., breaking when the right side is already too large. We can break from the c loop when $a^3 + b^3 < c^3$. (Surprisingly, doing an equivalent check on the d loop doesn't save us any time. Yes, we'd be breaking early from the innermost loop when d is very large. But, for all smaller values of d, we're running an extra several steps.)

We can also save a bit of time by removing duplicates. Consider the solutions below, all of which are essentially equivalent:

```
1   3³ + 60³   = 22³ + 59³
2   60³ + 3³   = 22³ + 59³
3   3³ + 60³   = 59³ + 22³
4   60³ + 3³   = 22³ + 59³
5   22³ + 59³  = 3³ + 60³
6   22³ + 59³  = 60³ + 3³
```

1

7   $59^3 + 22^3 = 3^3 + 60^3$
8   $22^3 + 59^3 = 60^3 + 3^3$

Only one of these needs to be printed.

We can cut out some of these duplicates by forcing a <= b and c <= d. This will prevent us from printing pairs which are equivalent other than a and b being swapped or c and d being swapped.

We still need to worry about the left and right side being swapped though. If we require a < c (this will be true for exactly one of the two sets), then we will remove this case too.

We can handle all of these by picking the appropriate start conditions of the for loops. If we start b at a, then b will always be greater than or equal to a. We can do the same thing for c, but starting it a + 1. (Why +1? Because if c = a, then d = b. The result $a^3 + b^3 = a^3 + b^3$ isn't very interesting.)

We can also conclude that c is not bigger than b. This is because a < b and c < d. If you consider the equation $a^3 + b^3 = c^3 + d^3$. It would be impossible for both of the left values to be less than both of the right values and still have the left and right sides be equal.

```
1   void cubesBetter(int max) {
2     for (int a = 0; a < max; a++) {
3       int acubed = a * a * a;
4       for (int b = a; b < max; b++) {
5         int bcubed = b * b * b;
6         for (int c = a + 1; c < b; c++) {
7           int ccubed = c * c * c;
8           if (acubed + bcubed < ccubed) break;
9           for (int d = c; d < max; d++) {
10            int dcubed = d * d * d;
11            if (acubed + bcubed == ccubed + dcubed) {
12              String sol = a + "," + b + "," + c + "," + d;
13              System.out.println(sol);
14            }
15          }
16        }
17      }
18    }
19  }
```

This helps, but the runtime will still be $O(N^4)$. We can do better.

Let's look at the equation we're given: $a^3 + b^3 = c^3 + d^3$. Once we've determined the values of a, b, and c, there's only one possible value for d. The only question is if that's an integer or not. So, rather than iterating through all possibilities for d, we can just check if the resulting d value is an integer.

```
1   void cubes(int max) {
2      for (int a = 0; a < max; a++) {
3         int acubed = a * a * a;
4         for (int b = a; b < max; b++) {
5            int bcubed = b * b * b;
6            for (int c = a + 1; c < b; c++) {
7               int ccubed = c * c * c;
8               if (acubed + bcubed < ccubed) break;
9
10              /* Compute cubed root of (a³ + b³ - c³) and check if
11               * it's an integer.*/
12              int d = (int) Math.round(
13                 Math.pow((acubed + bcubed - ccubed), 1.0 / 3.0));
14
15              if (d >= c && acubed + bcubed == ccubed + d * d * d) {
16                 String solution = a + "," + b + "," + c + "," + d;
17                 System.out.println(solution);
18              }
19           }
20        }
21     }
22  }
```

This is $O(N^3)$. This is better, but not yet optimal.

Let's think about what our algorithm does. The current approach is something like this:

```
for each pair (a, b) where a < 1000 and b < 1000:
    compute cubeAB = a³ + b³
    find pairs (c, d) that sum to cubeAB
```

For any given pair, we are iterating across all possible *other* pairs to see if they're equal.

Instead, we can just group these pairs by sum. This requires just iterating through the pairs one time.

As we iterate through the pairs, we create a mapping from sum -> pair (p, q). Then, we print out all pairs of pairs within each sum. That is, if we find $pair_1$, $pair_2$, $pair_3$, $pair_4$—each with a sum of x—we would print ($pair_1$, $pair_2$), ($pair_1$, $pair_3$), ($pair_1$, $pair_4$), ($pair_2$, $pair_3$), ($pair_2$, $pair_4$), and ($pair_3$, $pair_4$).

Our sum -> pairs mapping might look something like this:

```
260245440   = 82³ + 638³
            = 144³ + 636³
```

$$958595904 = 22^3 + 986^3$$
$$= 180^3 + 984^3$$
$$= 692^3 + 856^3$$

$$8587000 = 46^3 + 204^3$$
$$= 120^3 + 190^3$$

$$95880024 = 102^3 + 456^3$$
$$= 228^3 + 438^3$$

. . .

We can implement this with a hashtable, representing the pair (a, b) with a string for simplicity.

```
1   void cubes(int max) {
2      /* Compute pairs for each sum.*/
3      Hashtable<Integer, ArrayList<String>> map = computeSums(max);
4      /* Print pairs of pairs.*/
5      printSolutions(map);
6   }
7
8   /* Create map from each possible to sum to all pairs that give
9    * this sum.*/
10  Hashtable<Integer, ArrayList<String>> computeSums(int max) {
11     Hashtable<Integer, ArrayList<String>> sums =
12        new Hashtable<Integer, ArrayList<String>>();
13     for (int a = 0; a < max; a++) {
14        for (int b = a; b < max; b++) {
15           int sum = a * a * a + b * b * b;
16           String solution = a + "," + b;
17           /* Add sum -> pair to hashtable.*/
18           if (!sums.containsKey(sum)) {
19              sums.put(sum, new ArrayList<String>());
20           }
21           ArrayList<String> solutions = sums.get(sum);
22           solutions.add(solution);
23        }
24     }
25     return sums;
26  }
27
28  /* Print all pairs that sum to every value.*/
29  void printSolutions(Hashtable<Integer, ArrayList<String>> map) {
30     for (int sum : map.keySet()) {
31        ArrayList<String> solves = map.get(sum);
32        printSolutionsForSum(solves);
33     }
34  }
35
36  void printSolutionsForSum(ArrayList<String> solutions) {
37     for (int i = 0; i < solutions.size(); i++) {
```

```
38          for (int j = i + 1; j < solutions.size(); j++) {
39             String sol = solutions.get(i) + "," + solutions.get(j);
40             System.out.println(sol);
41          }
42       }
43 }
```

This solution takes $O(N^2)$ time where N is max size of a, b, c, and d.

16.20 Given a string, print all permutations of that string. You can assume the word does not have any duplicate characters.

pg 301

This is a classic recursion problem.

Let's approach this with an example that we build from the bottom up.

```
a   -> a
ab  -> ab, ba
abc -> abc, acb, bac, bca, cab, cba
```

How could we build all permutations of abcd off of any or all of these answers?

The main difference is the presence of d. If we have all permutations of abc, we could "splice" d into each of those strings (in all possible ways).

```
abc ->  abc, acb, bac, bca, cab, cba
           splice(abc, d) -> dabc adbc abdc abcd
           splice(acb, d) -> dacb adcb acdb acbd
           splice(bac, d) -> dbac bdac badc bacd
           splice(bca, d) -> dbca bdca bcda bcad
           splice(cab, d) -> dcab cdab cadb cabd
           splice(cba, d) -> dcba cdba cbda cbad
```

The code below implements this.

```
1  ArrayList<String> permutations(String word) {
2     /* Base case: word is empty.*/
3     if (word.length() == 0) {
4        ArrayList<String> list = new ArrayList<String>();
5        list.add(word);
6        return list;
7     }
8
9     /* Remove last char and get permutations of remainder.*/
10    String lastChar = word.substring(word.length() - 1);
11    String remainder = word.substring(0, word.length() - 1);
12    ArrayList<String> list = permutations(remainder);
13    ArrayList<String> result = new ArrayList<String>();
14
15    /* Go through all permutations of the substring, splicing
```

```
16      * lastChar into it.*/
17    for (String partial : list) {
18      /* Splice lastChar into all possible positions.*/
19      for (int i = 0; i < partial.length(); i++) {
20        String left = partial.substring(0, i);
21        String right = partial.substring(i);
22        String spliced = left + lastChar + right;
23        result.add(spliced);
24      }
25      result.add(partial + lastChar); // also splice into end
26    }
27    return result;
28  }
```

This algorithm will take O(N!) time (where N is the number of characters in the string) since there are N! permutations.

We can't optimize this algorithm, but there is another approach. This is less intuitive for many people.

If we have the string abcd, we can build it from subresults as follows:

```
perms(abcd) =   {a + perms(bcd)}
              + {b + perms(acd)}
              + {c + perms(abd)}
              + {d + perms(abc)}
```

That is, we remove each character and permute the remaining. Then, we prepend the removed character to each permutation.

Rather than prepending each character to its "subpermutations," we can let the subpermutation handle this. The permutations function gets a prefix string, which represents what currently needs to be prepended, and permutes the rest.

```
1   void permutations(String word, String prefix) {
2     /* Our prefix is fully built. Print it.*/
3     if (word.length() == 0) {
4       System.out.println(prefix);
5     }
6
7     /* Remove each character. Permute the remainder, passing
8      * along the prefix.*/
9     for (int i = 0; i < word.length(); i++) {
10      char c = word.charAt(i);
11      String left = word.substring(0, i);
12      String right = word.substring(i + 1);
13      permutations(left + right, c + prefix);
14    }
15  }
```

Like the earlier approach, this is O(N!).

16.21 *In a group of people, a person is called a "celebrity" if everyone knows them but they know no one else. You are given a function knows(a, b) which tells you if person a knows person b. Design an algorithm to find the celebrity (if one exists).*

For simplicity, you can assume that everyone is given a label from 0 to N-1. You need to implement a function int findCelebrity(int N).

Observe that:

(1) There can only be one celebrity at most (due to the definition of a celebrity).

(2) The knows function is the only way to look up who knows who.

pg 301

Let's start with a simple brute force approach. We can iterate through all possible people, checking if this person is a celebrity. As soon as we find a person who fits the criteria of being a celebrity, we can return this person.

```
1    int findCelebrity(int people) {
2        for (int i = 0; i < people; i++) {
3            if (isCelebrity(people, i)) {
4                return i;
5            }
6        }
7        return -1;
8    }
9
10   boolean isCelebrity(int people, int candidate) {
11       for (int i = 0; i < people; i++) {
12           if (i != candidate) {
13               if (knows(candidate, i) || !knows(i, candidate)) {
14                   return false;
15               }
16           }
17       }
18       return true;
19   }
```

This takes $O(N^2)$ time since we are potentially calling knows(a, b) on every pair of people.

Let's see if we can do this faster.

Consider a call to knows for two people, x and y. The result of knows(x, y) will either be true or false. What can we conclude from these results?

- Suppose knows(x, y) = true. This means that x knows y. In this case, we know that x is not a celebrity. Celebrities can't know anybody.

- Suppose knows(x, y) = `false`. This means that x does not know y. In this case, we know that y is not a celebrity since everyone must know the celebrity.

This gives us a very interesting learning: given two people who are both potential celebrities, we can always eliminate one person as the celebrity.

> Note: Now is a good time to pause to try to figure out the rest of the solution.

If we can always eliminate one person as the celebrity, then we should be able to trim down our list of N people to just one potential celebrity in N-1 calls to knows. At that point, we can then verify that candidate really is the celebrity (since there could be no celebrities) by calling knows again for candidate and every other person.

This can be done in two passes:

1. Find the candidate.

2. Verify that the candidate is the celebrity.

Before diving into the code, let's think about how we implement the first pass: finding the candidate.

We could have a list that we remove people from as they are eliminated. Shifting elements around in a list is time consuming and, frankly, more work than necessary.

If we imagine our calls to knows, we can think about our algorithm as kicking things off with 0 as the candidate. When we call knows(0, 1), we will either eliminate 0 or 1. If we eliminate 0, then candidate becomes 1. We then move on to knows(candidate, 2).

Step 2 uses the same isCelebrity method that we implemented earlier.

```
1   int findCelebrity(int people) {
2       int candidate = findCandidate(people);
3       if (isCelebrity(people, candidate)) {
4          return candidate;
5       }
6       return -1;
7   }
8
9   int findCandidate(int people) {
10      int candidate = 0;
```

```
11    for (int i = 0; i < people; i++) {
12       /* If candidate gets ruled out, move on to i.*/
13       if (knows(candidate, i)) {
14          candidate = i;
15       }
16    }
17    return candidate;
18 }
19
20 boolean isCelebrity(int people, int candidate) {
21    for (int i = 0; i < people; i++) {
22       if (i != candidate) {
23          if (knows(candidate, i) || !knows(i, candidate)) {
24             return false;
25          }
26       }
27    }
28    return true;
29 }
```

This algorithm takes $O(N)$ time.

16.22 *You have an NxN matrix of characters and a list of valid words (provided in any format you wish). A word can be formed by starting with any character and then moving up, down, left, or right. Words do not have to be in a straight line (PACKING is a word below). You cannot reuse a letter for the same word, so GOING (in the grid below) would not be a word since it reuses the G. Design an algorithm and write code to print all valid words.*

```
L I G O
E P N I
N A C K
S M A R
```

Let's think about this algorithm step by step. We need to find all words that start with each letter. We can just iterate through each letter on the grid, kicking off a search for words that start with each letter.

But how do we find all words that start with a particular letter, like P?

From P, we can move either up, down, left, or right. This means that we can think about all words that start with P in the grid as being:

```
all words that start with P =
          all words that start with PI
     +    all words that start with PE
     +    all words that start with PA
     +    all words that start with PN
```

This leads to a natural recursive algorithm. We recurse in each direction (up, down, left, and right), building a word as we go. Whenever we have a complete word, we print it and continue with the recursion.

One tricky part is how we prevent ourselves from reusing a letter for the same word. There are a number of solutions for this, but all take the same general approach of marking a character as being "in use" while we traverse down its path. Afterwards, we unmark it so we can use it again.

We have used a boolean array to do this. Before we traverse to a cell's neighbors, we mark this cell as being taken. After we're done, we mark it as available again.

We can also perform an optimization in short circuiting early in our recursion. Imagine we have built the string PNCKR. That's certainly not the start of any valid word in our dictionary, so why continue recursing down this path?

If we implement the dictionary as a trie, we can have a function that tells us if a string is a substring of a valid word in the dictionary. A trie is special type of tree generally used for storing lists of words. It gives very efficient runtime to call an isPrefix method.

For this algorithm, we can use isPrefix to terminate the recursion if we are on an invalid string.

```
1   /* Find all words on board by finding all words that start with
2    * each character on the board.*/
3   void boggle(char[][] board) {
4     boolean[][] marked =
5       new boolean[board.length][board[0].length];
6     for (int i = 0; i < board.length; i++) {
7       for (int j = 0; j < board[0].length; j++) {
8         boggle(board, i, j, "", marked);
9       }
10    }
11  }
12
13  /* Find all words that start with prefix and use the charater
14   * at row, col.*/
15  void boggle(char[][] board, int row, int col, String prefix,
16              boolean[][] marked) {
17    /* Check that char is on board and not currently in use.*/
18    if (!inBounds(board, row, col) || marked[row][col]) {
19      return;
20    }
21
22    /* Append character to current word.*/
23    prefix = prefix + board[row][col];
24
25    /* If there are no words starting with this prefix, return.*/
26    if (!isValidPrefix(prefix)) {
```

```
27        return;
28     }
29
30     /* Found a word. Print it.*/
31     if (isValidWord(prefix)) {
32        System.out.println(prefix);
33     }
34
35     /* Mark character as in use.*/
36     marked[row][col] = true;
37
38     /* Traverse each of its neighbors.*/
39     boggle(board, row - 1, col, prefix, marked); // Go up
40     boggle(board, row, col + 1, prefix, marked); // Go right
41     boggle(board, row + 1, col, prefix, marked); // Go down
42     boggle(board, row, col - 1, prefix, marked); // Go left
43
44     /* We are done traversing its neighbors and will now return
45      * to its parent. Mark this cell as available again.*/
46     marked[row][col] = false;
47  }
48
49  /* Check if row, col is in bounds.*/
50  boolean inBounds(char[][] board, int row, int col) {
51     if (row < 0 || col < 0 ||
52         row >= board.length || col >= board[row].length) {
53        return false;
54     }
55     return true;
56  }
```

Describing the runtime of this algorithm is a bit tricky because it really depends on what the board and the English language is like. If many paths are valid (that is, the words form valid prefixes), then it will be much slower than if a lot of paths are not valid.

If we didn't do the isPrefix check, we would traverse through N^2 characters, For each character, we would move in four possible directions the first time and three after that. A path could be as long as N^2 (the number of characters), so the number of all possible paths starting from a given character is $O(4*3^{(N^2)})$. This give us a time of $O(N^2*4*3^{(N^2)})$, which reduces to $O(N^2*3^{(N^2)})$.

Realistically, given the trie and the pattern of letters in the English language, it will be much faster than that.

16.23 *Given an array of integers (with both positive and negative values), find the contiguous sequence with the largest sum. Return just the sum.*

Example:

```
Input:   2, -8, 3, -2, 4, -10
Output:  5 (i.e., {3, -2, 4})
```

pg 302

Let's start off with a brute force and see how we can optimize it.

Brute Force

We could iterate through all possible subsequences, comparing their sum to a maximum sum. At the end, we return the biggest we have seen.

```
1   int getMaxSum(int[] a) {
2       int maxSum = 0;
3       for (int left = 0; left < a.length; left++) {
4           for (int right = left + 1; right < a.length; right++) {
5               int sum = 0;
6               /* Add all values between
7               for (int i = left; i <= right; i++) {
8                   sum += a[i];
9               }
10              if (sum > maxSum) {
11                  maxSum = sum;
12              }
13          }
14      }
15      return maxSum;
16  }
```

This is $O(N^3)$. We can do better!

Brute Force (Optimized)

Since each subsequence can be uniquely described with a start point and end point, we know that there are roughly $O(N^2)$ subsequences of an array. And yet, our earlier algorithm is taking $O(N^3)$ time. This suggests we might be able to optimize this.

Let's consider what the innermost for loop (the i loop) is doing. It's just computing the sum of all the elements between left and right. We have just finished computing (in the prior iteration of the right loop) the sum of everything between left and right - 1.

Instead of recomputing the sum every time, we can just keep a running sum.

When right goes to the next iteration, we just add a[right] to the running sum.

```
1   int getMaxSumBF(int[] a) {
2      int maxSum = 0;
3      for (int left = 0; left < a.length; left++) {
4         int runningSum = 0;
5         for (int right = left; right < a.length; right++) {
6            runningSum += a[right];
7            if (runningSum > maxSum) {
8               maxSum = runningSum;
9            }
10        }
11     }
12     return maxSum;
13  }
```

We're now down to $O(N^2)$ time. This is better, but we're still essentially doing a brute force solution.

Optimized

Let's inspect what this last solution did at a deeper level.

2, -4, 4, -3, 2, 5, -1, -4, -5, -2, -1, 2

We moved through all possible subsequences. That includes, for example, subsequences that include the first two values (2 and -4). Why would we ever want a subsequence that starts with {2, -4}? Their sum is -2, which means that they will only make a subsequence's sum smaller.

We *do* sometimes want negative values in the subsequence, but only when the negative value can join bigger values on both sides.

This leads us to a useful insight: whenever a subsequence is negative, we know we won't want to include it.

Let's fix up our code to break early when runningSum goes negative, so that we can then try the next value of left.

```
1   int getMaxSum(int[] a) {
2      int maxSum = 0;
3      for (int left = 0; left < a.length; left++) {
4         int runningSum = 0;
5         for (int right = left; right < a.length; right++) {
6            runningSum += a[right];
7            if (runningSum > maxSum) {
8               maxSum = runningSum;
9            } else if (runningSum < 0) {
10              break;
```

```
11              }
12          }
13          return maxSum;
14      }
15  }
```

With this change, we now break as soon as we get past {2, -4}. Left will move on to point to -4, and then to 4 after that.

We'll continue moving `right` until `runningSum` becomes negative. When `runningSum` is bigger than `maxSum`, we'll update `maxSum`.

When does `runningSum` become negative? Let's walk through it.

```
left:         4
right:        4 | -3 | 2 | 5 | -1 | -4 | -5 | ...
runningSum:   4 |  1 | 3 | 8 |  7 |  3 | -2 |
maxSum:                   8
```

We break when `right` is pointing to -5. We have now definitely found the largest subsequence that starts at `left`.

Observe that, up until that point, the sum of the values between `left` and any point x was greater than or equal to 0. In other words:

```
sum(array[left], array[left + 1], ..., array[x-1]) > 0
```

Imagine a subsequence starting at x and continue to anywhere in the array. If `sum(array[left], array[left + 1], ..., array[x-1]) > 0`, then any subsequence starting at x could be made larger by instead starting it at `left`. It is, therefore, less optimal to start at x.

Thus, we haven't just found the largest subsequence that starts at `left`. We've found the largest subsequence that starts *anywhere* between `left` and `right`.

We should now just move `left` over to `right + 1`.

This brings us to a new algorithm:

1. Start `left` and `right` at the far left side.

2. Move `right` until `runningSum` becomes negative.

   » Track `runningSum` and `maxSum` along the way.

3. When `runningSum` becomes negative, move left over to `right + 1` and reset `runningSum`.

```
1   int getMaxSum(int[] a) {
2       int maxSum = 0;
3       int runningSum = 0;
```

```
4      int left = 0;
5      for (int right = 0; right < a.length; right++) {
6         runningSum += a[right];
7         if (runningSum < 0) {
8            left = right + 1;
9            runningSum = 0;
10        }
11        if (maxSum < runningSum) {
12           maxSum = runningSum;
13        }
14     }
15     return maxSum;
16  }
```

If you look carefully, you might notice that left is set but never actually used. Therefore, we can implement the code without it.

```
1   int getMaxSum(int[] a) {
2      int maxSum = 0;
3      int runningSum = 0;
4      for (int right = 0; right < a.length; right++) {
5         runningSum += a[right];
6         if (maxSum < runningSum) {
7            maxSum = runningSum;
8         } else if (runningSum < 0) {
9            runningSum = 0;
10        }
11     }
12     return maxSum;
13  }
```

At this point, we know that we're done optimizing. This code runs in $O(N)$ time and computes the longest sequence in a single pass of the array. We can't do better than that.

Appendix

This appendix includes a variety of resources to help you in your preparation. For additional resources, or to discuss questions with fellow PMs, please check out CrackingThePMInterview.com.

Ian McAllister: Top 1% PMs vs. Top 10% PMs

Ian McAllister (@ianmcall) started and leads the AmazonSmile program. He manages product management, software development, and UX design teams, and also does business development. Previously, he ran Amazon's world-wide gifting business and worked as a program manager for Microsoft.

What distinguishes the top 1% of product managers from the top 10?

The top 10% of product managers excel at a few of these things. The top 1% excel at most or all of them:

- **Think big** - A 1% PM's thinking won't be constrained by the resources available to them today or today's market environment. They'll describe large disruptive opportunities, and develop concrete plans for how to take advantage of them.

- **Communicate** - A 1% PM can make a case that is impossible to refute or ignore. They'll use data appropriately, when available, but they'll also tap into other biases, beliefs, and triggers that can convince the powers that be to part with headcount, money, or other resources and then get out of the way.

- **Simplify** - A 1% PM knows how to get 80% of the value out of any feature or project with 20% of the effort. They do so repeatedly, launching more and achieving compounding effects for the product or business.

- **Prioritize** - A 1% PM knows how to sequence projects. They balance quick wins vs. platform investments appropriately. They balance offense and defense projects appropriately. Offense projects are ones that grow the business. Defense projects are ones that protect and remove drag on the business (operations, reducing technical debt, fixing bugs, etc.).

- **Forecast and measure** - A 1% PM is able to forecast the approximate benefit of a project, and can do so efficiently by applying past experience and leveraging comparable benchmarks. They also measure benefit once projects are launched, and factor those learnings into their future prioritization and forecasts.

- **Execute** - A 1% PM grinds it out. They do whatever is necessary to ship. They recognize no specific bounds to the scope of their role. As necessary, they recruit, they produce buttons, they do bizdev, they escalate, they tussle with internal counsel, they *.

- **Understand technical trade-offs** - A 1% PM does not need to have a CS degree. They do need to be able to roughly understand the technical complexity of the features they put on the backlog, without any costing input from devs. They should partner with devs to make the right technical trade-offs (i.e. compromise).

- **Understand good design** - A 1% PM doesn't have to be a designer, but they should appreciate great design and be able to distinguish it from good design. They should also be able to articulate the difference to their design counterparts, or at least articulate directions to pursue to go from good to great.

- **Write effective copy** - A 1% PM should be able to write concise copy that gets the job done. They should understand that each additional word they write dilutes the value of the previous ones. They should spend time and energy trying to find the perfect words for key copy (button labels, nav, calls-to-action, etc.), not just words that will suffice.

I'm not sure I've ever met a 1% PM, certainly not one that I identified as such prior to hiring. Instead of trying to hire one, you're better off trying to hire a 10% PM who strives to develop and improve along these dimensions.

This essay originally appeared on http://www.quora.com/Product-Management/ What-distinguishes-the-Top-1-of-Product-Managers-from-the-Top-10/answer/Ian-McAllister.

Adam Nash: Be a Great Product Leader

Adam Nash is the Chief Operating Officer at Wealthfront. Before Wealthfront, Adam served as an Executive in Residence at Greylock Partners, where he advised the leadership teams of the firm's existing consumer technology companies as well as evaluating new investment opportunities. Prior to joining Greylock, Adam was Vice President of Product Management at LinkedIn.

People who know me professionally know that I'm passionate about Product Management. I truly believe that, done properly, a strong product leader acts as a force multiplier that can help a cross-functional team of great technologies and designers do their best work.

Unfortunately, the job description of a product manager tends to either be overly vague (you are responsible for the product) or overly specific (you write product specifications). Neither, as it turns out, is it effective in helping people become great product managers.

I've spent a lot of time trying to figure out a way to communicate the value of a product manager in a way that both transparently tells cross-functional partners what they should expect (or demand) from their product leaders, and also communicates to new product managers what the actual expectations of their job are. Over the years, I reduced that communication to just three sets of responsibilities: Strategy, Prioritization & Execution.

Responsibility #1: Product Strategy

They teach entire courses on strategy at top tier business schools. I doubt, however, that you'll hear Product Strategy discussed in this way in any of them.

Quite simply, it's the product manager's job to articulate two simple things:

* What game are we playing?

* How do we keep score?

Do these two things right, and all of a sudden a collection of brilliant individual contributors with talents in engineering, operations, quality, design and marketing will start running in the same direction. Without it, no amount of prioritization or execution management will save you. Building great software requires a variety of talents, and key innovative ideas can come from anywhere. Clearly describing the game your playing and the metrics you use to judge success allows the team, independent of the product manager, to sort through different ideas and decide which ones are worth acting on.

Clearly defining what game you are playing includes your vision for the product, the value you provide your customer, and your differentiated advantage over competitors. More important, however, is that it clearly articulates the way that your team is going to win in the market. Assuming you pick your metrics appropriately, everyone on the team should have a clear idea of what winning means.

You should be able to ask any product manager who has been on the job for two weeks these questions, and get not just a crisp, but a compelling answer to these two questions.

The result: aligned effort, better motivation, innovative ideas, and products that move the needle.

Responsibility #2: Prioritization

Once the team knows what game they are playing and how to keep score, it tends to make prioritization much easier. This is the second set of responsibilities for a product manager—ensuring that their initial work on their strategy and metrics is carried through to the phasing of projects / features to work on.

At any company with great talent, there will be a surplus of good ideas. This actually doesn't get better with scale, because as you add more people to a company they tend to bring even more ideas about what is and isn't possible. As a result, brutal prioritization is a fact of life.

The question isn't what is the best list of ideas you can come up with for the business—the question is what are the next three things the team is going to execute on and nail.

Phasing is a crucial part of any entrepreneurial endeavor—most products and companies fail not for lack of great ideas, but based on mistaking which ones are critical to execute on first and which can wait until later.

Personally, I don't believe linear prioritization is effective in the long term. I've written a separate post on product prioritization called The Three Buckets that explains the process that I advocate.

You should be able to ask any product manager who has been on the job for two weeks for a prioritized list of the projects their team is working on, with a clear rationale for prioritization that the entire team understands and supports.

Responsibility #3: Execution

Product managers, in practice, actually do hundreds of different things.

In the end, product managers ship, and that means that product managers cover whatever gaps in the process that need to be covered. Sometimes they author content. Sometimes they cover holes in design. Sometimes they are QA. Sometimes they do PR. Anything that needs to be done to make the product successful they do, within the limits of human capability.

However, there are parts of execution that are massively important to the team, and without them, execution becomes extremely inefficient:

- **Product specification** – the necessary level of detail to ensure clarity about what the team is building.

- **Edge case decisions** – very often, unexpected and complicated edge cases come up. Typically, the product manager is on the line to quickly triage those decisions for potentially ramifications to other parts of the product.

- **Project management** – there are always expectations for time/benefit trade-offs with any feature. A lot of these calls end up being forced during a production cycle, and the product manager has to be a couple steps ahead of potential issues to ensure that the final product strikes the right balance of time to market and success in the market.

- **Analytics** – in the end, the team largely depends on the product manager to have run the numbers, and have the detail on what pieces of the feature are critical to hitting the goals for the feature. They also expect the product manager to have a deep understanding of the performance of existing features (and competitor features), if any.

Make Things Happen

In the end, great product managers make things happen. Reliably, and without fail, you can always tell when you've added a great product manager to a team versus a mediocre one, because very quickly things start happening. Bug fixes and feature fixes start shipping. Crisp analysis of the data appears. Projects are re-prioritized. And within short order, the key numbers start moving up and to the right.

Be a great product leader.

This essay originally appeared on http://blog.adamnash.com/2011/12/16/be-a-great-product-leader/.

Sachin Rekhi: The Inputs to a Great Product Roadmap

Sachin Rekhi is a serial entrepreneur with a product management background. He founded Connected (acquired by LinkedIn), Feedera (acquired by LinkedIn), and Anywhere.fm (acquired by imeem). He is now a group product manager at LinkedIn and has also worked as a program manager for Microsoft.

I'm often asked how I think about coming up with the product roadmap for an upcoming release. To help answer this, I thought I'd share how my team recently went about thinking through the roadmap for an upcoming product we're working on.

Analysis of existing usage metrics

When you're innovating on top of an existing product, the best place to start is by conducting an in-depth analysis of the existing usage patterns of your product. Understanding at a high level the features that are most used will tell you where further investment may be justified. Low usage features also give you insights into what might need a redesign or need to be removed altogether. Diving into flow analysis also helps you understand what optimizations could be made to reduce friction in the current experience.

User interviews to understand audience pain points

Great products provide solutions to great problems, so it's always important to ensure you're solving the problems most top of mind for your target audience. The best way I've found to really understand these pain points is through user interviews seeking to understand your audience's motivations, daily workflow, existing tools, current frustrations, and more. The focus here should be problem space more than solution space to really get at what problems would warrant product solutions.

Aggregation of customer feedback & support requests

Users are constantly reaching out to creators of products with feature suggestions, support requests, complaints, and more. It's valuable to spend the time to aggregate this feedback to understand trends amongst your user base and what areas might be worth investing in.

In-depth look at competition

Taking a look at other players in the space to see what's working for them and what isn't is another great source of product ideas for your roadmap. Using the products, reviewing the product's user forums, and reading product and industry reviews is a great way to uncover what's most interesting about your competition. While it's important to play your own game compared to your competition, it's nonetheless an important source of input to consider.

Commercialization of internal innovation

Oftentimes your product is a part of the overall suite of offerings your company provides. Each product tends to innovate on their own dimensions and bring to market what's most important for their audience and product area. Nonetheless oftentimes there is a significant opportunity to bring similar innovations to your target audience or product area. I thus find it helpful to stay abreast of the latest releases from other company products to see what might be leveragable in your own product area.

Audience surveys to understand feature prioritization

Once you've contemplated a set of potential features, it's often helpful to survey a portion of your existing or potential users to help prioritize these features against each other. Leveraging conjoint analysis can help get at the relative importance of each of your features to better understand which are worth investing in.

These various inputs help inform your product roadmap by helping you discover key themes across the various inputs that may be critical areas to attack in your next product release. However, it's important to keep in mind that developing the right product roadmap remains as much art as it is science. While these inputs can help inform your potential roadmap, it's the creative synthesis of these that ultimately result in a great roadmap.

This essay originally appeared at http://www.sachinrekhi.com/blog/2013/09/23/ the-inputs-to-a-great-product-roadmap.

Ken Norton: How to Hire a Product Manager

Ken Norton is a partner at Google Ventures. Before that, he was a group product manager at Google. He joined Google in 2006 with the acquisition of JotSpot, where he was vice president of product. Before that, he was senior director of product management at Yahoo.

It's been a while since I was hiring at a startup, and recruiting at a startup is very different from hiring at a big company. At Yahoo! Search, it seemed like we were constantly hiring. I did an average of 5-8 interviews a week. It was a never-ending drumbeat of resumes, interviews, and offer letters. Now, I wasn't always the hiring manager. I only hired a handful of product managers in my time there. But somebody was always hiring a product manager and I was usually on the interview team. The first thing you notice at a big company is the amount of specialization. At a startup, everyone does a little of everything, so you need strong generalists. More importantly, it's hard to predict the future, so you need people who can adapt. You might think you're hiring somebody to work on something specific, but that something might change in a few months. It doesn't work that way at big companies. Usually when you're hiring you have a very specific role in mind, and the likelihood that that responsibility will change is low. Lots of people were hired at Yahoo! that probably wouldn't have been appropriate at a startup. I recall a lot of post-interview conversations that went something like this - "well, I'm not sure they're the perfect candidate, but they do seem suited for this very specific role, so let's hire them." That may work fine at a big company, but it's deadly thinking at a startup.

I started my career as an engineer and advanced pretty quickly into engineering management. During the bubble, I probably hired over one hundred engineers. I learned a lot about hiring, mostly by making mistakes. When I transitioned to product management I was able to apply some of my experience hiring technical people, but I also learned a whole new set of lessons. Last week a friend called to say he needed to hire a product manager and wanted my advice. I realized there's not a lot of good information out there about interviewing PMs (there's not a lot of good information about product management in general). More to the point, there's not a lot about what you should look for in a product manager no matter what kind of environment you're in - startup or big company. So I thought I'd pull together some of what I learned.

Remember buddy, nobody asked you to show up

Product management may be the one job that the organization would get along fine without (at least for a good while). Without engineers, nothing would get built. Without sales people, nothing is sold. Without designers, the product looks like crap. But in a world without PMs, everyone simply fills in the gap and goes on with their lives. It's important to remember that - as a PM, you're expendable. Now, in the long run great product management usually makes the difference between winning and losing, but you have to prove it. Product management also combines elements of lots of other specialties - engineering, design, marketing, sales, business development. Product management is a weird discipline full of oddballs and rejects that never quite fit in anywhere else. For my part, I loved the technical challenges of engineering but despised the coding. I liked solving problems, but I hated having other people tell me what to do. I wanted to be a part of the strategic decisions, I wanted to own the product. Marketing appealed to my creativity, but I knew I'd dislike being too far away from the technology. Engineers respected me, but knew my heart was elsewhere and generally thought I was too "marketing-ish." People like me naturally gravitate to product management.

1. Hire all the smart people

So what do I look for in a PM? Most importantly, raw intellectual horsepower. I'll take a wickedly smart, inexperienced PM over one of average intellect and years of experience any day. Product management is fundamentally about thinking on your feet, staying one step ahead of your competitors, and being able to project yourself into the minds of your colleagues and your customers. I usually ask an interview candidate a series of analytical questions to gauge intelligence and problem-solving ability. Generally I'll ask questions until I'm sure the candidate is smarter than me. For some reason, lots of people I know are reluctant to do that. They argue that it's insulting to the candidate. I think the right candidate will relish the challenge. In fact, that's the first test - how do they react when I say "I'd like to pose some theoretical problems, is that okay?" The best of the bunch are usually bouncing out of their chairs with excitement. The super smart sometimes counter with questions of their own.

2. Strong technical background

Some managers I've known insist on hiring only PMs with computer science degrees. I'm not as snobby - maybe it's my own liberal arts undergraduate education - but I do tend to favor people who've been in technical roles. Having a solid engineering background gives a PM two critical tools - the ability to relate to engineers and a grasp of the technical details driving the product. It depends on the product of course - a PM working on low-level developer APIs is bound to need more technical chops than one working on the front-end of a personals web site. But the basic principle applies - product managers with technical backgrounds will have more success conveying product requirements to engineers and relaying complicated details to non-technical colleagues and customers. That said, there are pitfalls you need to avoid. Most importantly, a PM who's a former engineer needs to realize that he or she is just that - a former engineer.

PMs who come from engineering and still try to take charge of technical decisions and implementation details will crash spectacularly. For that reason, I like hiring technical people who've already made the move to product management at a previous job. They've already gone through the challenging adaptation period and by checking references you can get a feel for how well they've evolved. I won't bore with you with interview questions to evaluate technical competency. They depend on the skill set and there are hundreds of web sites that give good tips for hiring engineers. Instead, here are some good questions for gauging how well a technical PM has adapted to the role and their ability to work with engineers:

- Why did you decide to move from engineering to product management?

- What is the biggest advantage of having a technical background?

- What is the biggest disadvantage?

- What was the biggest lesson you learned when you moved from engineering to product management?

- What do you wish you'd known when you were an engineer?

- How do you earn the respect of the engineering team?

3. "Spidey-sense" product instincts and creativity

This next category is highly subjective, difficult to evaluate, and extraordinarily important. I am a strong believer that certain people are born with innate product instincts. These people just know what makes a great product. They're not always right, but their instincts usually point in the right direction. They tend to be passionate advocates of a point of view, sometimes to the chagrin of their colleagues. I've had the good fortune to work with a good number of these people, and it's an essential trait in product managers. And it can be tuned, but it can't be learned. Product management, especially in highly dynamic environments like the web, involves lots of small decisions. Sure, there's a lot of big thinking and strategy. But it's the little decisions where a great PM distances him or herself from a decent one. You know they've got the "spidey-sense" product instinct when they suggest approaches that nobody on the team has thought of, but immediately strike everyone as obvious when they hear them. Evaluating product instinct in an interview is challenging at best. But it can be done. One thing I always do is check to see if the candidate has accomplished the following tasks during a one-hour interview:

- **Independently echoed some of my own concerns about my product** - if you're a good PM, you've got a bunch of things that worry you about your own product. Maybe they're UI shortcomings, missing features, or architecture flaws that need to be addressed. They're things you know need to be fixed. At least some of these should be obvious to an intelligent outsider with strong product instincts. I look for that moment in the interview when I smile, nod, and say "yeah, I know - that's been driving us crazy too."

- **Taught me something new about my product** - it could be an obvious improvement that I'd never considered, a new idea for positioning against a competitor, or a problem they encountered that needs to be addressed. When I learn something from a candidate, I know two things: (1) they're not afraid to speak critically, and (2) they're probably smarter than me. I want both in a product manager.

- **Turned me on to something new and interesting** - people with great product instincts tend to notice great products before everyone else. If I'm interviewing a top-notch candidate, I usually walk away having discovered something new and innovative.

Here are some good questions for judging product instincts:

- Tell me about a great product you've encountered recently. Why do you like it? [By the way, it drives me crazy when candidates name one of my products in an interview. I had a hard time hiring anybody at Yahoo! who told me the coolest product they'd come across recently was Yahoo! Good grief.]

- What's made [insert product here] successful? [I usually pick a popular product, like the iPod or eBay, that's won over consumers handily in a crowded market.]

- What do you dislike about my product? How would you improve it?

- What problems are we going to encounter in a year? Two years? Ten years?

- How do you know a product is well designed?

- What's one of the best ideas you've ever had?

- What is one of the worst?

- How do you know when to cut corners to get a product out the door?

- What lessons have you learned about user interface design?

- How do you decide what not to build?

- What was your biggest product mistake?

- What aspects of product management do you find the least interesting and why?

- Do you consider yourself creative?

4. Leadership that's earned

Product managers are usually leaders in their organizations. But they typically don't have direct line authority over others. That means they earn their authority and lead by influence. Leadership and interpersonal skills are critical for product management. There are a thousand books about leadership, so I won't turn this post into a treatise on the subject (most of the books are crap anyway). I find reference checks to be the most effective way to measure leadership skills, especially references that involve peers and individual contributors who worked with - but did not report to - the candidate. But here are a few questions I've used in the past:

- Is consensus always a good thing?

- What's the difference between management and leadership?

- What kinds of people do you like to work with?

- What types of people have you found it difficult to work with?

- Tell me about a time when a team didn't gel. Why do you think that happened, and what have you learned?

- How do you get a team to commit to a schedule?

- What would somebody do to lose your confidence?

- Do you manage people from different functions differently? If so, how?

- What have you learned about saying no?

- Who has the ultimate accountability for shipping a product?

- Have you ever been in a situation where your team has let you down and you've had to take the blame?

- How has your tolerance for mistakes changed over the years?

- Which do you like first, the good news or the bad news?

- What's your approach to hiring?

5. Ability to channel multiple points-of-view

Being a product manager requires wearing multiple hats. I often joke that much of the time your job is to be the advocate for whoever isn't currently in the room - the customer, engineering, sales, executives, marketing. That means you need to be capable of doing other people's jobs, but smart enough to know not to. Great PMs know how to channel different points-of-view. They play devil's advocate a lot. They tend to be unsatisfied with simple answers. In one conversation they might tell you the requirements don't seem technically feasible and in the next breath ask how any of this will make sense to the salespeople. There's one obvious way to evaluate a candidate's ability to think through a problem from multiple angles - gets lots of people in the interview process. I always insist that at a minimum, representatives from engineering, design, and marketing meet a potential PM candidate. Depending on the specific role, this list can grow - pre-sales engineering, support, developer relations, business development, legal, or customers themselves. Ultimately anyone who will be working with this person should meet them. Note that I didn't say everyone needs to meet them. One carefully selected representative of each key function will suffice. And it also doesn't mean everybody has to give a thumbs-up - it's hard to build consensus in an interview process as the list of interviewers grows, so consider the feedback appropriately. But nobody will be able to judge how well a product manager understands the sales process like a salesperson. I also strongly recommend that you give specific instructions to the interviewers, like "I'd like you to see how well this person would understand the issues you face in channel development, and how we'll they'd support you in the field. "Here are some specific questions that I use (these are just examples, feel free to replace the functional names):

• How have you learned to work with sales?

• What is the best way to interface with customers?

• What makes marketing tick?

• How do you know when design is on the right track?

• How should a product manager support business development?

• What have you learned about managing up?

• What's the best way to work with the executives?

6. Give me someone who's shipped something

This last characteristic may be the easiest to evaluate. Unless the position is very junior, I'll usually hire product managers who've actually shipped a product. I mean from start to finish, concept to launch. Nothing is a better indication of someone's ability to ship great products than having done it before. Past performance is an indication of future success. Even better, it gives something tangible to evaluate in a sea of intangibles. When checking references, I always make sure to talk to important colleagues from a previous project, especially the PM's manager and their engineering and sales or marketing counterparts. (Incidentally, these rules are ordered for a reason, and as I mentioned under #1 I'll still take a brilliantly smart PM over a dimmer experienced one even if the former hasn't shipped before).

Note: I wrote this in 2005 when I was at JotSpot. Google acquired JotSpot in 2006. Since then, I've had the opportunity to work with some marvelous PMs and have performed 200+ PM interviews. I'm sure that my opinions have evolved, but the intervening years have only further reinforced the characteristics of great PMs. I occasionally set out to update this essay but I always decide to leave it as is.

This essay originally appeared on https://www.kennethnorton.com/essays/productmanager.html.

Amazon Leadership Principles

The following leadership principles are reprinted from amazon.com. If you are applying for a job at Amazon, you should definitely read them. They might be useful for other companies as well, since many companies look for similar attributes.

Whether you are an individual contributor or the manager of a large team, you are an Amazon leader. These are our leadership principles and every Amazonian is guided by these principles.

Customer Obsession

Leaders start with the customer and work backwards. They work vigorously to earn and keep customer trust. Although leaders pay attention to competitors, they obsess over customers.

Ownership

Leaders are owners. They think long term and don't sacrifice long-term value for short-term results. They act on behalf of the entire company, beyond just their own team. They never say "that's not my job."

Invent and Simplify

Leaders expect and require innovation and invention from their teams and always find ways to simplify. They are externally aware, look for new ideas from everywhere, and are not limited by "not invented here." As we do new things, we accept that we may be misunderstood for long periods of time.

Are Right, A Lot

Leaders are right a lot. They have strong business judgment and good instincts.

Hire and Develop the Best

Leaders raise the performance bar with every hire and promotion. They recognize exceptional talent, and willingly move them throughout the organization. Leaders develop leaders and take seriously their role in coaching others.

Insist on the Highest Standards

Leaders have relentlessly high standards - many people may think these standards are unreasonably high. Leaders are continually raising the bar and driving their teams to deliver high quality products, services and processes. Leaders ensure that defects do not get sent down the line and that problems are fixed so they stay fixed.

Think Big

Thinking small is a self-fulfilling prophecy. Leaders create and communicate a bold direction that inspires results. They think differently and look around corners for ways to serve customers.

Bias for Action

Speed matters in business. Many decisions and actions are reversible and do not need extensive study. We value calculated risk taking.

Frugality

We try not to spend money on things that don't matter to customers. Frugality breeds resourcefulness, self-sufficiency, and invention. There are no extra points for headcount, budget size, or fixed expense.

Vocally Self Critical

Leaders do not believe their or their team's body odor smells of perfume. Leaders come forward with problems or information, even when doing so is awkward or embarrassing. Leaders benchmark themselves and their teams against the best.

Earn Trust of Others

Leaders are sincerely open-minded, genuinely listen, and are willing to examine their strongest convictions with humility.

Dive Deep

Leaders operate at all levels, stay connected to the details, and audit frequently. No task is beneath them.

Have Backbone; Disagree and Commit

Leaders are obligated to respectfully challenge decisions when they disagree, even when doing so is uncomfortable or exhausting. Leaders have conviction and are tenacious. They do not compromise for the sake of social cohesion. Once a decision is determined, they commit wholly.

Deliver Results

Leaders focus on the key inputs for their business and deliver them with the right quality and in a timely fashion. Despite setbacks, they rise to the occasion and never settle.

Acknowledgements

Thank you to everyone who has supported me throughout this journey.

To Paul Unterberg, Adam Kazwell, and our two anonymous candidates: I deeply appreciate your being willing to put yourself out there with your resumes. Thank you.

To Fernando Delgado, Ashley Carroll, Brandon Bray, Thomas Arend, Johanna Wright, Lisa Kostova Ogata, Ian McAllister, Adam Nash, Sachin Rekhi, and Ken Norton: Your contributions are much appreciated by me, Jackie, and our readers.

To all my PM (and related) friends and colleagues who contributed their advice and feedback: You all are awesome.

To my husband John and his mother Donna: Thank you for your continuous support which made this book possible.

To my beautiful son, Davis: One day you'll be big enough to read this, and you'll know how much you're loved.

GAYLE

Thank you to my loving family. My father's dissatisfaction with the long lines at toll bridges became my first PM-interview-question practice. My husband encouraged me throughout the book writing process. My cats always had a helping paw.

I'm immensely grateful to Steven Sinofsky and Marissa Mayer who hired me into their talented PM orgs and have advanced the craft of product management. I'm also very thankful to my wonderful managers: Mike Morton, Tom Stocky, Jack Menzel, Johanna Wright, and Justin Rosenstein who have been amazing mentors and role models.

Thank you to all of the people who talked to me about their experiences as PMs and shared their advice. I was blown away by how many of you shared my vision for this book and helped it come into being. I'd especially like to thank Shirin Oskooi for helping me get things started, Daniel Dulitz for suggesting the section on transitioning from designer to PM, Nundu Janakiram for suggesting the section on PM myths, and Chrix Finne for connecting me to lots of helpful people.

JACKIE

Gayle Laakmann McDowell

Gayle Laakmann McDowell is the founder / CEO of CareerCup.com, a site dedicated to preparing for tech jobs.

She has performed hundreds of interviews as an employee at Google, coached numerous software engineering and product manager candidates, and helped many startups through their dev and PM acquisition interviews with top tech companies.

Prior to CareerCup, she worked for Microsoft, Apple, and Google. Most recently, Gayle spent three years at Google as a software engineer, where she was one of the company's lead interviewers and served on Google's hiring committee.

This is Gayle's third book. Her second book, The Google Resume, is a compre- hensive book offering advice on how anyone can prepare for a role at a top tech company. Her first book, Cracking the Coding Interview, is a deep dive into coding interviews and is Amazon's best-selling interview book.

Gayle holds a bachelor's and master's degree in Computer Science from the University of Pennsylvania and an MBA from the Wharton School.

She lives in Palo Alto, California.

facebook.com/gayle
twitter.com/gayle
technologywoman.com
quora.com/Gayle-Laakmann-McDowell

Jackie Bavaro

Jackie Bavaro is a product manager at Asana, a leading startup that builds the modern productivity software for teamwork without email used by companies such as Dropbox, Airbnb, Uber, Foursquare, and Pinterest. She joined as the company's first product manager and she lead the team through their public launch, the launch of their premium product, and the launch of their product for larger teams.

With over eight years of experience in product management, she has also worked as a product manager at Google and as a program manager at Microsoft. At Google, she joined as part of the elite Associate Product Manager (APM) program and worked on Google Search, where notably she launched Place Search - the first product to group web results around objects in the real world. At Microsoft, she worked in the Office group on Windows SharePoint Server where she launched blogs and wikis on the SharePoint platform.

Jackie has interviewed over 100 PM candidates in both phone screens and on-site interviews, reviewed many resumes, and sourced numerous candidates. She has advised many candidates on applying to become product managers and finding the right company, which was part of the inspiration for writing this book.

More of Jackie's writing about product management can be found on The Art of Product Management at http://pmblog.quora.com where she delves into topics like communication, building relationships with the team, and making data-driven decisions.

Jackie graduated from Cornell University with a double major in Computer Science and Economics.

She lives in San Francisco, California.

facebook.com/jackie.bavaro
twitter.com/jackiebo
pmblog.quora.com
quora.com/Jackie-Bavaro

good luck, folks.

Made in the USA
Lexington, KY
09 May 2015